COMPUTATIONAL MORPHOLOGY

Machine Intelligence and Pattern Recognition

Volume 6

Series Editors

L. N. KANAL

and

A. ROSENFELD

University of Maryland
College Park
Maryland
U.S.A.

NORTH-HOLLAND
AMSTERDAM · NEW YORK · OXFORD · TOKYO

Computational Morphology

A Computational Geometric Approach to the Analysis of Form

Edited by

Godfried T. TOUSSAINT
School of Computer Science
McGill University
Montreal, Quebec
Canada

N·H
P C

1988

NORTH-HOLLAND
AMSTERDAM · NEW YORK · OXFORD · TOKYO

ISBN: 0 444 70467 1

Publishers:
ELSEVIER SCIENCE PUBLISHERS B.V.
P.O. Box 1991
1000 BZ Amsterdam
The Netherlands

Sole distributors of the U.S.A. and Canada:
ELSEVIER SCIENCE PUBLISHING COMPANY, INC.
52 Vanderbilt Avenue
New York, N.Y. 10017
U.S.A.

LIBRARY OF CONGRESS
Library of Congress Cataloging-in-Publication Data

Computational morphology : a computational geometric approach to the analysis of form / edited by Godfried T. Toussaint.
p. cm. -- (Machine intelligence and pattern recognition ; v. 6)
Includes index.
ISBN 0-444-70467-1 (U.S.)
1. Geometry--Data processing. 2. Convex domains. I. Toussaint, Godfried T., 1944- . II. Series.
QA448.D38C66 1988
516--dc19 88-17467
CIP

PRINTED IN THE NETHERLANDS

To Guylaine Lefebvre

PREFACE

A word is in order concerning the title of this book – *Computational Morphology*. Morphology is the study of *form*. The form of an object is the *shape* and *structure* of that object as distinguished from its material components. Thus morphology is concerned with describing the shape and structure of objects. As such it is a central concern of many disciplines. For example, it is a branch of biology that deals with the changing shape and structure of evolving animals and plants. In linguistics it is the study and description of word formation in a language. Visual perception in psychology is concerned with how the visual systems of humans and other animals and organisms recognize shapes and patterns. Computer vision researchers attempt to program computers to recognize or describe scenes of objects according to their shape. Geometry is a branch of mathematics that deals with the description of and interrelationships among objects such as points, lines, surfaces and solids. While the shape of an object may enter into such a description it does so serendipidously. The *shape* or *form* of an object is not a well formalized concept as is a purely geometric structure such as for example the *convex hull* of an object. Shape and form are *morphological* rather than *geometrical* and are determined as much by internal, subjective, perceptions as by external, objective descriptions of the perceptual stimulus. This is not to say that one cannot use formal quantitative methods to analyse, describe and synthesize shape and form. Indeed, one of the earliest attempts at obtaining a mathematical analysis of shape is the seminal work of D'Arcy Thompson dating back to 1917, *On Growth and Form*, Cambridge University Press. Presently, mathematical morphology is a flowering discipline (see for example *Image Analysis and Mathematical Morphology*, by J. Serra, Academic Press, 1982). However, any mathematical description of shape and form thus far conceived, no matter how sophisticated, has always fallen short of what was hoped for.

Computational geometry is a maturing discipline of computer science that in a strong sense was born about fifteen years ago. It is concerned with the design and analysis of algorithms for solving geometric problems. It has found application in many areas such as statistics, computer graphics, robotics, architecture, geographical analysis, operations research, image processing, pattern recognition and computer vision, for example. Since morphology is intimately related to geometry it is not surprising that computational geometry should be able to make a large contribution to the science of form. When a *computational geometric* structure or operator is intended to extract (measure) the *shape* or *form* of an object to which it is applied, it is referred to as a problem in *computational morphology*. A typical problem then in computational morphology is: given a set of points S in the plane, what is the "shape" of S? Constructing the *convex hull* of S is a computational geometric solution, albeit a poor one, to such a problem. We see from this example that the boundary between *computational-geometry* and *computational-morphology* is necessarily fuzzy. Nevertheless it is not difficult to isolate those problems in computational geometry that have direct bearing on computational morphology.

This book contains a collection of fourteen papers covering a wide variety of topics in *computational morphology* relevant to such disciplines as morphogenesis, morphometry, stereology, image processing, pattern recognition, geographical analysis, cluster analysis and computational perception. Some of the papers are of a tutorial-survey nature while others present the latest research results. The collection should serve well as a graduate text for courses in computational geometry, pattern recognition, image processing and computer vision, as well as provide researchers interested in the science-of-form with some of the latest developments.

One of the most frequent ways in which an object is described is in the structure of a polygon. A key step in obtaining a morphological description of a polygon is its decomposition into simpler components, i.e., components that are either convex, star-shaped, or spirals. Although researchers have investigated such decompositions extensively, little effort has been devoted towards understanding their inherent complexity. A. Aggarwal, S.K. Ghosh and R.K. Shyamasundar investigate the computational complexity of a class of polygon decomposition techniques in which the boundaries of the resulting components are composed of edges which are segments of lines passing through any two vertices of the given polygon.

In many applications such as graphics and image processing the polygons are often *rectilinear* or *isothetic*, i.e., their edges are parallel to either coordinate axes. J. Sack and G. Toussaint consider the problem of decomposing such polygons into star-shaped components. They present a linear time algorithm for the case of monotone polygons and an $0(n \log \log n)$ time algorithm for arbitrary polygons of size n. This problem has a guard-placement interpretation. Since a star-shaped polygon is visible from a single point (guard) the entire polygon can be guarded by placing one guard in each star-shaped component. Their algorithm finds a decomposition into at most $[n/4]$ star-shaped components.

One can divide the types of polygon decompositions into: (1) those that attempt to solve explicitly the *morphological* problem of realizing decompositions into perceptually meaningful components (also called "anthropomorphic" in the literature) and, (2) those that decompose polygons into "arbitrary" components such as triangles, convex quadrilaterals and so on. Of course an "arbitrary" decomposition into say, star-shaped, polygons may in fact yield a quite satisfactory decomposition from the morphological point of view. One decomposition which is "procedure oriented" rather than "component oriented" and which yields perceptually meaningful components is the relative-neighbour decomposition. The previous best algorithm for this problem had a worst-case complexity of $0(n^3)$. Here Hossam ElGindy and Godfried Toussaint show how this decomposition can be computed in $0(n^2)$ time.

Another morphological problem which is related to the question "What is the shape of a set of points?" is: given two sets of points S_1 and S_2 of n points each, how "separated" are they? This question is of course also central to the design of decision rules for the automatic classification of shapes. Once d measurements have been made on a shape it can be considered as a point in a d-dimensional Euclidean space. If we have a set of data consisting of n shapes from one class and n from another then knowing how "separated" they are is very useful in designing the simplest possible decision rule. Binay Bhattacharya presents an algorithm for computing all the circles that separate two given sets S_1 and S_2 of points in the plane in $0(n \log n)$ time. The previous fastest algorithm for this problem is due to S. Fisk and runs in $0(n^2)$ time.

It is a well-known result in computational geometry that given a set of n points in the plane one can always draw a simple polygon with those points as the vertices of the polygon. In fact this can easily be done in $0(n \log n)$ time. Take the average of any three of the points as a center 0. Sort all the points by polar angle about 0 and connect them in this order. The resulting polygon is not only simple but also star-shaped from 0. The morphological version of this "connect-the-dots" game asks for drawing a simple polygon through a set of points with the added constraint that the polygon should reflect the perceptual shape of the set of points. Joesph O'Rourke considers the problem of connecting a set of dots into a collection of *rectilinear* polygons.

One can also play the game of "connect-the-line-segments". Given a set of n non-intersecting line segments in the plane can one draw a perceptually meaningful simple polygon through the set of line segments. Surprisingly, unlike the case of "connect-the-dots" this is not always possible with line segments. David Avis and David Rappaport present polynomial-time algorithms for deciding whether a set of line segments admits a *monotonic* polygon.

When a shape is represented by a polygon which has been obtained from an image processing device there are often irregularities in the polygon due to noise during the image acquisition phase of the process. Researchers have therefore developed scores of algorithms for "smoothing" polygons, i.e., approximating polygons with polygons of fewer edges that still manage to capture the original shape of the polygon. Hiroshi Imai and Masao Iri present a tutorial survey of this area as well as an elegant, formal, graph-theoretical framework for discussing the approaches taken which leads to several efficient algorithms. One of these elegant methods originally took $0(n^3)$ time. Avraham Melkman and Joseph O'Rourke show how this complexity can be reduced to $0(n^2 \log n)$.

One of the main features of an object that contributes to a morphological description useful for shape analysis and classification in its symmetry. Several algorithms have appeared recently for detecting symmetries in objects. Peter Eades surveys results in this area and describes open problems for further research.

The prototypical problem in computational morphology is of course finding efficient algorithms for computing the "shape" of a planar set of points. John Radke reports on recent progress on the application of a new methodology for describing the "internal structure" of planar point sets to the analysis of both point set and network patterns.

Convexity has always played a fundametnal role in any discussion about shape. Classical convexity however is limited. Therefore researchers have sought to relax convexity in different directions. Star-shapedness and monotonicity are two examples of the relaxation of classical convexity. Recently the notions of geodesic-convexity and link-convexity have made a strong impact in several disciplines. See for example the paper "Computing the Link Center of a Simple Polygon", by Bill Lenhart, Ricky Pollack, Jorg Sack, Raimund Seidel, Micha Sharir, Subhash Suri, Godfried Toussaint, Sue Whitesides and Chee Yap (Proc. of Third ACM Symposium on Computational Geometry, Waterloo, Canada, June 1987, pp. 1-10). Yet another way to relax convexity is to consider only a fixed number of orientations in which the classical notions of convexity should hold. This topic is discussed in a tutorial fashion in the paper by Gregory Rawlins and Derick Wood.

One of the central problems in the morphological analysis of objects in three-dimensional space is their reconstruction from two-dimensional line-drawings. The realizability problem asks for whether a given two-dimensional line-drawing truly represents a real three-dimensional polyhedron. Kokichi Sugihara presents a tutorial survey of the realizability problem.

Traditionally computational geometry has worked in the "vector-graphics" mode where the objects being considered were either sets of points (each point represented by its cartesian coordinates) or polygonal regions where edges of the polygons are represented by the cartesian coordinates of their end-points. However, computational geometry can also of course be done in the "raster-graphics" mode which has been the preferred mode of working for researchers in image processing, computer vision and mathematical morphology. Which approach is more likely to be preferred in time only time can tell. In the meantime there is much to be gained by both camps through cooperation. Jun-ichiro Toriwaki and Shigeki Yokoi show how some recent results from "vector-graphics" computational morphology can very successfully be translated to "raster-graphics" algorithms for texture analysis.

The central problem in computer vision begins with a grey-level intensity array of visual pattern or scene and culminates with a description which is a marriage of the array itself (the bottom-up or data-driven information) and the purpose for which it is viewed (the top-down or conceptually-driven information). Considerable attention has been devoted to low-level vision, i.e., the aspect dealing with the analysis of the intensity array up to the level of figure-ground separation. The main purpose of analyzing an intensity array at this level is to form a low-level description that is independent of, or holds true for, any final conceptually-driven description. David Marr has called such a description a *primal sketch* of the intensity array. One class of patterns or scenes which have been studied extensively is the so-called *dot pattern*. In the last paper in this book, Godfried Toussaint proposes a new graph termed the *sphere-of-influence* graph as a *primal sketch* intended to capture the low-level perceptual structure of visual scenes consisting of dot patterns. This graph suffers from none of the serious drawbacks of previous methods and for a pattern consisting of n dots, can be computed efficiently in $0(n \log n)$ time.

I would like to express my sincere thanks to all the authors for their contribution, to the reviewers for their constructive comments, and a special thanks to Laveen N. Kanal for inviting me to put this volume together.

Godfried Toussaint
Montreal
February 5, 1988

TABLE OF CONTENTS

Computational Morphology
G.T. Toussaint (editor)

Computational Complexity of Restricted Polygon Decompositions

Alok Aggarwal
IBM Thomas J. Watson Research Center
Box 218, Yorktown Heights, NY 10598

S. K. Ghosh
Tata Institute of Fundamental Research
Homi Bhabha Road, Bombay 400005, India

and

R. K. Shyamasundar[1]
IBM Thomas J. Watson Research Center
Box 218, Yorktown Heights, NY 10598.

Abstract: This note investigates the computational complexity of a special kind of decomposition of polygons into convex, star-shaped, and spiral components. In these polygon decompositions, the boundary of the resulting component is composed of edges which are segments of lines passing through any two vertices of the polygon and the resulting components are allowed to overlap each other. We point out that for certain kinds of components (such as star-shaped), the problem of finding the minimum number of, possibly, overlapping pieces remains NP-hard even under this restriction and we provide an approximate algorithm that yields a solution that is guaranteed to be at most $O(\log n)$ times the optimal. We also provide an application of our approximate algorithm to guarding art galleries with almost the minimum number of watchmen.

Index Terms: pattern recognition, polygon decompositions, art gallery problem, convex, star-shaped, spiral-shaped, covers, NP-complete, approximation algorithms.

1. Introduction

This paper evaluates the complexity of a special kind of decomposition of polygons into convex, star-shaped, and spiral components. In this polygon decomposition, the boundary of the resulting component is composed of edges which are segments of lines passing through any two vertices of the given polygon and the resulting components are allowed to overlap each other so that they form a cover of the polygon. Figure 1 depicts an example of a restricted decomposition of the letter "E" into four convex pieces. In [FP75], it is argued by Feng and Pavlidis that the decomposition of polygons obtained by extending the edges alone is a very "natural" form of decomposition and they have investigated the problem of decomposing the polygon using edge extensions when the the resulting components form spiral polygons. In [FP75], Feng and Pavlidis also

[1] This work was done during Sept.-Dec. 1984 when this author was on leave from the Tata Institute of Fundamental Research, Homi Bhabha Road, Bombay 400005, India.

provide applications of such restricted decompositions in syntactic pattern recognition. Similarly, other researchers have shown that such restricted decompositions of a polygon into star-shaped and convex components are also "elegant" and they are widely applicable in character recognition and in the understanding of biomedical and aerial photographs. For example, in [Pa68], Pavlidis has argued that "restricted convex decompositions" provides an "elegant" and an "intuitive" description of a given polygon whereas Marayuma [Ma72] and Avis and Toussaint [AT81] have investigated star-shaped decompositions. For an excellent survey of various syntactic pattern recognition techniques and their applications, see [Pa78].

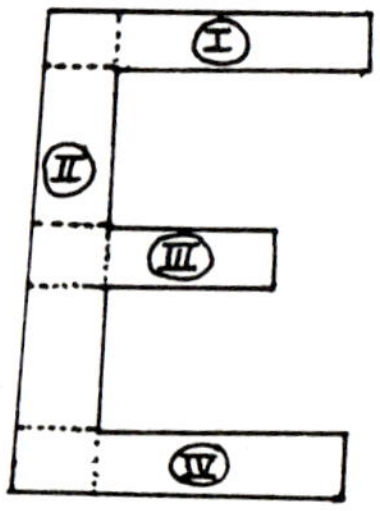

Figure 1

Although researchers have investigated such decompositions, their efforts have been mainly directed towards the development of decomposition heuristics for pattern recognition. In particular, very little effort has been made towards the understanding of the computational aspects of such decompositions. This note attempts to mitigate this deficiency by exploring some complexity issues involved in such decompositions. It is clear that the class of decompositions where the components are bounded by edge extensions alone is a subset of the class in which we allow the components to be bounded by segments of lines passing through any two vertices of the polygon. In the next two sections, we consider decompositions using only edge extensions and, in the final section, we generalize these results to decompositions obtained by segments of lines passing through any two vertices of the polygon.

2. Restricted Convex and Spiral Decompositions

Consider the problem of convex decomposition of a simple polygon when the resulting components are bounded only by the segments of its edge extensions. Following [O'R82a], let $v_1, \ldots, v_n$ be the clockwise chain of vertices on the boundary of a polygon where each $v_i = (x_i, y_i)$. Let $h_i(v_i, v_{i+1})$ define the right half plane obtained by extending the edge connecting v_i to v_{i+1}. If $e_1, \ldots, e_n$ are the edges of the polygon such that edge e_i connects v_i to v_{i+1}, then these edges define the corresponding half planes $h_1, \ldots, h_n$. These half planes are referred to as the positive half planes of the polygon and they consist of regions that are to the right of the directed lines containing $e_1, \ldots, e_n$. Let $\bar{h}_1, \ldots, \bar{h}_n$ represent the half planes obtained by reversing the direction of the edges so that $\bar{h}_i$ represents the right-half plane corresponding to the edge that connects v_{i+1} to v_i. Then, $\bar{h}_i$ is referred to as the negative half plane of the polygon. The notion of *primary convex set* is defined by Pavlidis [Pa68] is given as follows: A *primary convex set* A is formed by the intersection of a subset of n *positive* half planes of the polygon, P, such that the following conditions hold:

1. $A \subseteq P$.
2. There is no set B which is obtained as the intersection of some subset of the positive half planes of P such that $B \subseteq P$ and $A \subseteq B$.

In [Pa68], Pavlidis shows that every convex subset of P which is not a primary convex subset is always contained in some primary convex set (Corollary 1, page 168). Consequently, Pavlidis restricts his attention to covering the polygon with primary convex sets only and shows that for any given polygon the maximum number of such sets is at most $\binom{n}{n-3}$, i. e., $O(n^3)$. However, a critical analysis of the proof of Theorem 1 given in [Pa68] shows that Corollary 1 is incorrect. The following example illustrates this fact by showing that at times a primary convex set may not contain a non-primary convex set. Consider the nine-sided simple polygon shown in Figure 2. This polygon can be covered with the following three convex subsets which are represented by the intersection of their half planes:

1. $P_1 = h_1 \cap \bar{h}_7 \cap h_8 \cap h_9$
2. $P_2 = h_2 \cap h_3 \cap h_4$
3. $P_3 = h_5 \cap h_6 \cap h_7$

Clearly, P_1 is not a primary convex set and a simple analysis shows that at least four primary convex sets are required to cover the given polygon. Consequently, the minimum cover may sometimes contain non-primary convex sets. Define a *maximal convex set* as one that is not contained in any other convex set and that is formed by the intersection of both positive and negative half planes. Observe that any convex subset of P whose every edge is a segment of a line going through two vertices of P and that is not maximal is always contained in some maximal convex subset of P. Consequently, unlike the situation for primary convex sets, we can, indeed, restrict our attention to covering the polygon with maximal convex subsets. However, also, unlike primary convex sets, we show below that the number of maximal convex sets may be as large as $2^{n/3}$ for some n-sided polygons.

To demonstrate a polygon with $3k$ sides that has 2^k maximal convex sets, consider a regular convex k-gon and let edges e_i and $e_{(i+1) \bmod k}$ be incident on vertex v_i for $0 \le i \le k-1$. Then, we obtain the required polygon by replacing v_i by a pair of edges (e_i', e_i'') such that the extension of e_i' intersects $e_{(i+1) \bmod k}$ and the extension of e_i'' intersects e_i. Figures 3, 4, and 5 depict

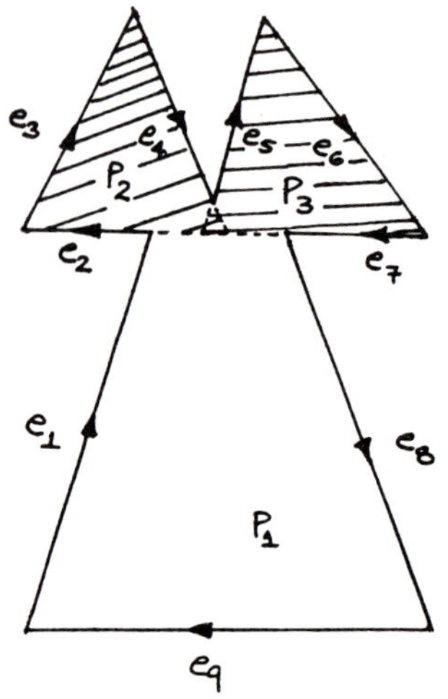

Figure 2

polygons with $k = 2, 3$, and 4, respectively. This construction demonstrates that if, for the restricted class of minimum convex cover, an algorithm enumerates all maximal sets then its execution requires exponential time. Similarly, if we are interested in obtaining restricted minimum spiral decompositions, then we can generalize the notion of maximal convex sets to maximal spiral sets and it is not hard to construct $9k$-sided polygons which can be decomposed into 2^k maximal spirals.

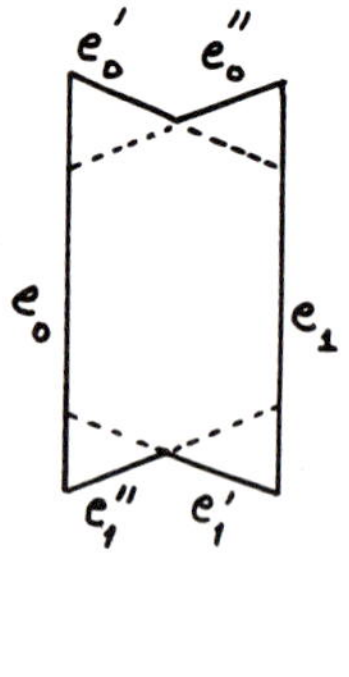

Figure 3

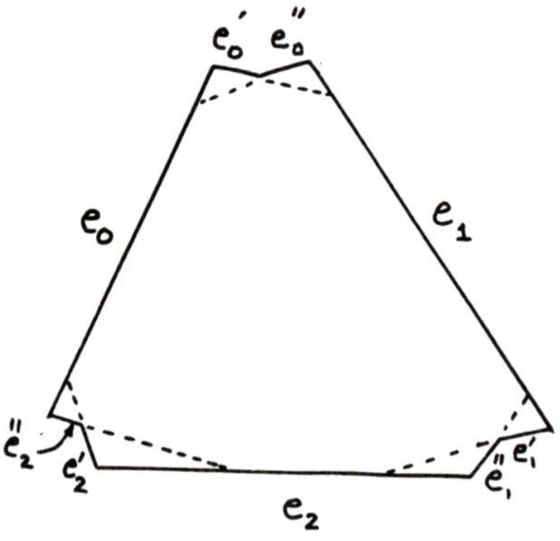

Figure 4

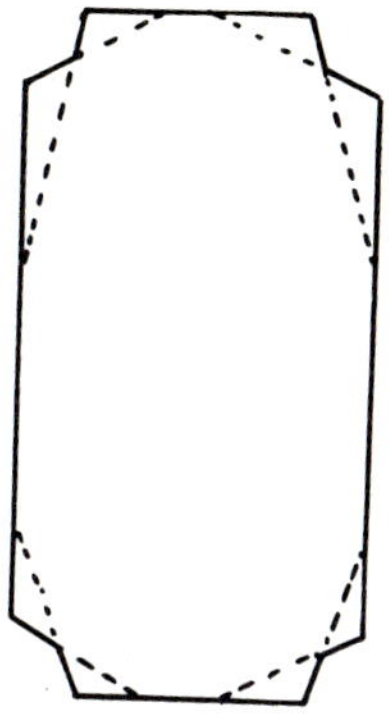

Figure 5

3. Computing Restricted Star-Shaped Decompositions

As in section 2, let the edges of the resulting components in a star-decomposition be obtained only as segments of the extensions of the polygonal edges. Furthermore, let C_{opt} be the minimum number of star-regions in any star cover of *P* where the regions are bounded by such edge extensions. In this section, we show that unlike convex and spiral covers, the number of star regions that need to be considered for any minimum star cover is bounded by $\binom{n}{2}$, i. e., by $O(n^2)$. We use this to demonstrate an $O(n^4 \log n)$ algorithm which computes a star cover of an *n*-sided polygon that consists of at most $O(\log n \times C_{opt})$ star regions. For the description of this algorithm and for the proof of sufficiency of $O(n^2)$ star regions, we need the following definitions:

Let *x* be any point that belongs to *P*. Then, *x* is said to see another point $y \in P$ if the segment *xy* does not properly intersect the boundary of *P*. The visibility region $VIS(x)$ is the set of all points that are visible to *x* and it is easy to show that $VIS(x)$ forms a simple polygon. ElGindy and Avis [EA81] and Lee [Le83] have independently obtained linear time algorithms for computing the visibility polygon for any point that belongs to *P*. The kernel of a polygon is a region such that any point in this region can see the entire polygon. We define the *restricted visibility polygon* with respect to *x*, $RVIS(x)$, as follows:

1. $RVIS(x)$ is bounded by the edge extensions of *P*.
2. If a point $y \in RVIS(x)$, then *x* sees *y*.
3. There is no other polygon that follows conditions (1) and (2) and that contains $RVIS(x)$.

A convex set *a* is called a *basic cycle* if *a* has non-zero area and if *a* is the intersection of some (positive or negative) half planes and the following conditions hold:

1. $a \subseteq P$.
2. There is no convex set *b* which is the intersection of some (positive or negative) half planes such that $b \subseteq a$.

Observe that we do not allow the possibility of a basic cycle to be a degenerate point or a line segment; a basic cycle is the dual of a maximal convex set in that it is the minimal non-empty convex set that is formed by the intersection of some subset of $2n$ half planes. Since there are n edge extensions in P, a straightforward analysis shows that the total number of intersections of these edges is at most $\binom{n}{2}$ so that there are at most $O(n^2)$ basic cycles. The utility of basic cycles in computing star-covers is demonstrated in the following Proposition:

Proposition 1: Let $Q_1, Q_2, \ldots, Q_C$ form a star cover of P such that for $1 \leq i \leq C$, the edges of Q_i belong to segments of the edge extensions of P. Then, the kernel of Q_i contains at least one basic cycle of P.

Proof: For any i between 1 and C, direct all edges of Q_i in the clockwise direction and define positive half planes with respect to the directed edges. Since Q_i is a star-shaped polygon, its kernel, K_i, is non-empty, and a simple analysis shows that K_i can be obtained as the intersection of the positive half planes of Q_i. Now, since Q_i is totally contained in P, clearly, K_i cannot contain or intersect any edge of P. Also, since the edges of Q_i belong to the edge extensions of P, it follows that K_i is obtained as the intersection of some (positive or negative) half planes of P. Consequently, if K_i is a basic cycle in P then the Proposition follows trivially; otherwise, it must contain some basic cycle of P which establishes this Proposition.

■

The importance of Proposition 1 is obvious: for restricted minimum star-covers, we only need to consider basic cycles as possible candidates for the kernels of the star-regions. Since there are $O(n^2)$ basic cycles, the computational requirements of any algorithm which computes the restricted star-cover would generally be lower than a corresponding algorithm that computes a restricted convex cover. Let $J_1, J_2, \ldots, J_m$ denote the basic cycles of P. Then, clearly, P can be partitioned into $J_1, J_2, \ldots, J_m$. Furthermore, if $Q_1, Q_2, \ldots, Q_C$ are as defined in Proposition 1 then it is easy to see that for every Q_i, there is a set of J_j's such that each of the J_j's is entirely contained in Q_i and Q_i can be partitioned into these J_j's. Below, we use these observations to provide an $O(n^4 \log n)$-time approximation algorithm, *APPROX-STAR*, which computes a star cover with no more than $O(\log n \times C_{opt})$ components where C_{opt} denotes the minimum number of restricted star-shaped components that are required to cover P.

Algorithm APPROX-STAR

Step 1: Given the n-sided polygon P, compute the basic cycles of P and denote these by $J = \{J_1, J_2, \ldots, J_m\}$.

Step 2: For $1 \leq i \leq m$, let u_i be some vertex of J_i. Then, determine the restricted visibility polygon, $RVIS(u_i)$, and note that $RVIS(u_i)$ can be partitioned into some basic cycles of P. Let $RVIS(u_i)$ be composed of the basic cycles: $\{J_{1,i}, J_{2,i}, \ldots, J_{k_i,i}\} \subseteq J$.

Step 3: Draw a bipartite graph, G, with partitions $X = \{x_1, x_2 \ldots x_m\}$ and $Y = \{y_1, y_2 \ldots y_m\}$ such that (x_i, y_j) is an edge in G if $u_i \in J_i$ and $RVIS(u_i)$ contains the basic cycle J_j. Let $X' \subseteq X$ be the set with minimum cardinality such that every vertex in Y is adjacent to at least one vertex in X'. Then, we will show in Theorem 1 that $C_{opt} = |X'|$. So, if we can find X' then we have obtained the restricted, minimal cardinality star-shaped cover; however, problem of finding X' is the classical unweighted set-cover problem [GJ79] and it is known to be NP-hard. Consequently, we use the greedy approximate algorithm given by Chvatal [Ch79] that finds a set $X'' \subseteq X$ such that $|X''| = O(\log m \times |X'|) \leq O(\log n \times C_{opt})$ since $m \leq n^2$ and $|X'| = C_{opt}$.

Step 4: If $x_p \in X''$ then find $u_p \in J_p$ and $RVIS(u_p)$ and note that $\bigcup_{u_p \in J_p} RVIS(u_p)$ forms a restricted star-cover of P.

end;

Theorem 1: Algorithm *APPROX-STAR* correctly computes a restricted star-cover of P whose cardinality is at most $O(\log n \times C_{opt})$ and this algorithm can be executed in $O(n^4 \log n)$ time.

Proof: *APPROX-STAR* computes X'' and Chvatal's greedy algorithm guarantee its cardinality to be at most $O(\log n \times |X'|)$. Consequently, to prove the correctness of *APPROX-STAR*, we only need to show that $|X'| = C_{opt}$. Since every vertex of Y is adjacent to some vertex in X', it follows that every basic cycle belongs to some $RVIS(u_i)$ where $u_i \in J_i$ and J_i corresponds to some $x_i \in X'$. Consequently, such restricted visibility regions together form a restricted star-cover of P and hence $|X'| \geq C_{opt}$. On the other hand, because of Proposition 1, we simply have to look at the basic cycles for a restricted star-cover of P which implies that $C_{opt} \geq |X'|$ and it proves the correctness of *APPROX-STAR*.

To establish the time complexity, note that there are at most n edge extensions so the number of basic cycles $m \leq O(n^2)$. Consequently, Step 1 takes $O(n^3)$ time steps. Steps 2 and 4 each take at most $O(m \times n^2)$ time steps because for $1 \leq i \leq m$, if $u_i \in J_i$ then $RVIS(u_i)$ and S_i can be computed in $O(n^2)$ time. Finally, the greedy approximate algorithm given by Chvatal [Ch79] for finding the minimum set-cover takes $O(\log n \times$ the number of edges in $G)$ and since G has at most m^2 edges, Step 4, and hence the entire algorithm can be executed in $O(n^4 \log n)$ time steps.

■

4. Discussion

In sections 2 and 3, while computing restricted decompositions of a polygon into convex, spiral, or star-shaped components, we assumed that the resulting pieces are bounded by edge extensions of the given polygon. However, Feng and Pavlidis [FP75] have noted that if we consider segments that connect any two vertices to be the bounding edges, then the resulting pieces in spiral decompositions are quite "natural" for pattern recognition applications. Furthermore, the number of such pieces may often be less than those that are obtained by edge extensions alone. Figure 6

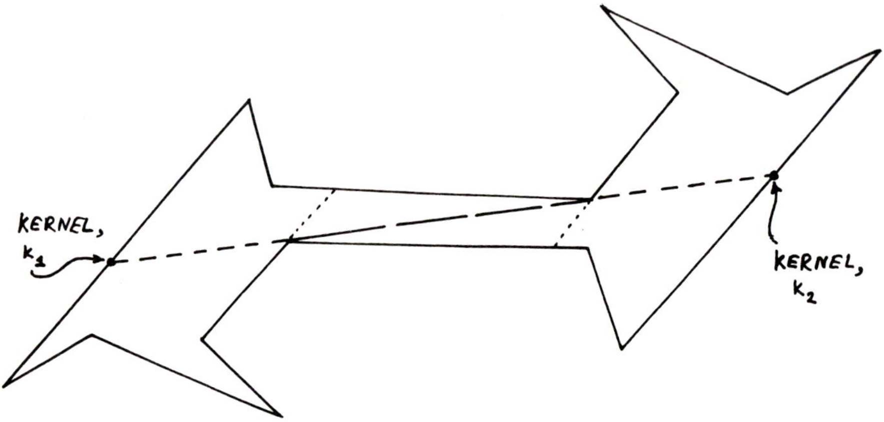

Figure 6

demonstrates that this is also true for star-shaped decompositions: the polygon in this figure requires at least three star pieces to cover it if only edge extensions are allowed whereas two star pieces suffice if we allow the edges to be bounded by segments that contain any two vertices of the polygon. Note that the arguments of Sections 2 and 3 can also be generalized for such decompositions. Since maximal convex and spiral sets can be obtained by incorporating all half planes induced by lines containing any two vertices of the polygon, the 6-vertex polygon that is given in Figure 3 still admits four maximal convex sets. Consequently, we can use the strategy given in section 2 to construct n-sided polygons that admit $2^{n/3}$ maximal convex or $2^{n/9}$ maximal spiral sets. In a similar manner, the notion of a basic cycle can be generalized for star-decompositions and the algorithm APPROX-STAR can be modified to compute $O(\log n \times C_{opt})$ star-regions. In this case, it is easy to see that number of basic cycles is $O(n^4)$ which implies that the time-complexity of APPROX-STAR is $O(n^8 \log n)$.

Consider the decompositions that allow segments of lines passing through any two vertices of the polygon. In such decompositions, although there are polygons that admit at least $2^{n/9}$ maximal convex and spiral pieces, it remains to determine whether finding the minimum number of restricted convex or spiral convex covers is NP-hard. Similarly, for other kinds of decompositions such as convex, spiral and star-shaped, where the resulting components are bounded by edge extensions of P, the complexity of finding minimum covers remains to be determined. Although, the complexity of most of these problems remains unknown, as far as star-decompositions are concerned, using a recent result of Lee and Lin [LL84], it can be shown that at least one of these decomposition problems is NP-hard.

Claim: Given a polygon $P = (v_1, ..., v_n)$ and a positive integer k, determining whether k restricted star-shaped pieces are sufficient to cover P is NP-complete where it is assumed that for a restricted star-shaped component, every edge of the component is segment of a line containing any two vertices of P.

The proof of this Claim follows in a straight forward manner from the proof given by Lee and Lin [LL84] and that given by Aggarwal [Ag84, O'R87]; since these proofs are rather tedious, we omit them here.

Although the restriction on the edges of the resulting components, as discussed above, is a very natural restriction, O'Rourke [O'R82b] has shown that, in fact, it is a true restriction. In other words, there are polygons whose minimum convex covers use pieces that have edges which do not form segments of lines containing any two vertices of P (see Figure 7). It remains to determine whether O'Rourke's examples can be generalized to show that even for spiral and star-shaped covers, our assumed restriction is indeed a true restriction. In a similar vein, a geometric characterization of unrestricted decompositions, from the viewpoint of obtaining fast algorithms for minimum covers, also remains to be determined.

During the entire paper, we restricted our attention to simple polygons without holes. Since

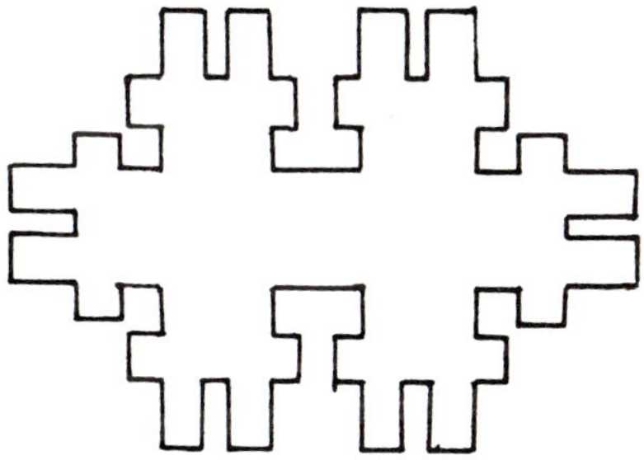

Figure 7(a) Adapted from [O'R82b]

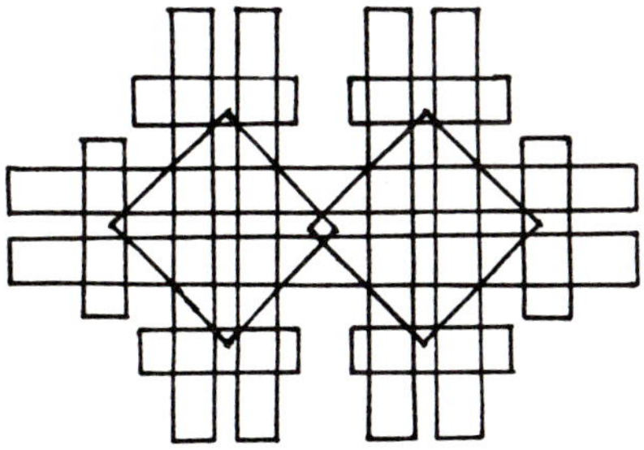

Figure 7(b) Adapted from [O'R82b]

the characters that have to be recognized in pattern recognition often have holes, it is important to study the complexity of restricted decompositions for polygons with holes. In this regard, the results of O'Rourke and Supowit [OS83] can be applied to establish that the problem of finding the minimum number of restricted convex, restricted spiral, or restricted star-shaped pieces is NP-hard. Again, although approximate algorithms for obtaining the minimum number of restricted convex or restricted spiral-shaped pieces remains unknown, algorithm *APPROX-STAR* can be easily extended to obtain $O(\log n \times C_{opt})$ restricted star-shaped pieces for a polygon that has holes, that has a total of n vertices, and that requires C_{opt} restricted star-pieces. Also, this algorithm can be executed in $O(n^8 \log n)$ time even for a polygon with holes.

Finally, we provide an application of *APPROX-STAR* to the art gallery problem. The art gallery problem is to determine the minimum number of guards that are sufficient to cover (or see every point in the interior of) an art gallery room. The room can be thought of as a simple polygon P with n vertices and the guards as stationary points in P. Usually the guards may be placed anywhere inside the polygon but if they are placed at the vertices, they are called *vertex guards*. Lee and Lin [LL84] have shown that even the problem of finding the minimum number of vertex guards for an art gallery is NP-complete. Although there are no known approximate algorithms for finding the minimum number of guards, *APPROX-STAR* can be easily modified so that it can be executed in $O(n^8 \log n)$ time, it obtains at most $O(\log n \times C_{opt})$ vertex guards for a polygon that may have holes, that has a total of n vertices, and that requires C_{opt} vertex guards. This modification is possible because all the bounding edges of the simply connected region that a vertex guard can see actually form segments of the lines going through two of the n vertices which implies that a vertex guard can only see a restricted star-shaped region.

Acknowledgments: The authors thank Joseph O'Rourke of the Johns Hopkins University for stimulating discussions. They also thank an anonymous referee for useful suggestions.

References

[Ag84] A. Aggarwal, "The art-gallery theorem: its variations, extensions, and algorithmic aspects," Ph. D. Thesis, the Johns Hopkins University, Baltimore, Maryland, 1984.

[AT81] D. Avis and G. T. Toussaint, "An efficient algorithm for decomposing a polygon into star-shaped polygons," Pattern Recognition, Vol. 13, 1981, pp. 295-398.

[Ch79] V. Chvatal, "A greedy-heuristic for the set covering problem," Math. Operations Res., Vol. 4, No. 3, 1979, pp. 233-235.

[EA81] H. Elgindy and D. Avis, "A linear time algorithm for computing the visibility polygon from a point," J. of Algorithms, Vol. 2, 1981, pp. 186-197.

[FP75] H-Y. F. Feng and T. Pavlidis, "Decomposition of polygons into simpler components: feature generation for syntactic pattern recognition," IEEE Trans. Comp., Vol. C-24, 1975, pp. 636-650.

[GJ79] M. R. Garey and D. S. Johnson, "Computers and Intractability: a guide to the theory of NP-completeness," W.H. Freeman and Company, 1979.

[Le83] D. T. Lee, "Visibility of a simple polygon," Computer Vision, Graphics and Image Processing, Vol. 22, 1983, pp. 207-221.

[LL84] D. T. Lee and A. K. Lin, " Computational complexity of art gallery problems," Technical Report, Northwestern University, 1984.

[Ma72] K. Maruyama, " A study of visual shape perception," Dept. of Computer Science, University of Illinois, TR UIUCDCS-R-72-533, 1972.

[O'R82a] J. O'Rourke, "Polygon decomposition and switching function minimization," Computer Graphics and Image Processing, Vol. 18, 1982, pp. 382-391.

[O'R82b] J. O'Rourke, "The complexity of computing minimum convex covers for polygons," Proc. Twentieth Allerton Conf. on Communications, Control and Computing, 1982, pp. 75-84.

[O'R87] J. O'Rourke, "Art Gallery Theorems and Algorithms," International Series of Monographs on Computer Science, Oxford University Press, New York, 1987.

[OS83] J. O'Rourke and K. J. Supowit, "Some NP-hard polygon decomposition problems," IEEE Trans. on Information Theory, Vol. IT-29, 1983, pp. 181-190.

[Pa68] T. Pavlidis, "Analysis of set patterns," Pattern Recognition, Vol. 1, 1968, pp. 165-178.

[Pa78] T. Pavlidis, "Review of algorithms for shape analysis," Computer Graphics and Image Processing, Vol. 7, 1978, pp. 243-258.

Computational Morphology
G.T. Toussaint (editor)

COMPUTING MONOTONE SIMPLE CIRCUITS IN THE PLANE

David Avis

School of Computer Science
McGill University
805 Sherbrooke St. West
Montréal, Québec, H3A 2K6

David Rappaport

Department of Computing and Information Science
Queen's University
Kingston, Ontario, K7L 3N6

Abstract: Given a set of line segments in the plane it is not always possible to form a simple circuit (simple polygon) by connecting the endpoints of the segments. In general, it is NP-complete to decide whether a set of line segments admits a simple circuit. We give polynomial algorithms to decide whether a set of line segments admits a monotone simple circuit. We also give polynomial algorithms for optimizing the perimeter or the area bounded by the simple circuit. The algorithms presented are based on a dynamic programming approach for solving a special case of the travelling salesman problem.

1. INTRODUCTION

A natural generalization of the problem of finding simple circuits through a set of points is the problem of finding simple circuits through a set of line segments. The task of obtaining a simple circuit through a set of points is a recurring theme that appears in a variety of applications. In problems that involve routes through networks, the tour of shortest Euclidean length that begins and ends at a common site is a simple circuit. This is the Euclidean travelling salesman problem, and is known to be NP-complete [3, 5]. Simple circuits have also been used in the area of pattern recognition for extracting perceptual information from sparse data [7,11,12]. Surprisingly, the problem of computing simple circuits through a set of line segments has not received much attention.

The problem of computing simple circuits from line segments was brought to the attention of the authors by Godfried Toussaint. Fáry [2] showed that every simple polygon can be triangulated. A simple way to prove that every set of points can be triangulated is to first show that every set of points admits a simple circuit that bounds a simple polygon. The interior and exterior of this polygon is then triangulated. Following this reasoning one could also show that a set of disjoint line segments could be triangulated by showing that every set of disjoint line segments admits a simple circuit. In [1] it is shown that indeed every set of line segments can be triangulated. However, not every set of disjoint line segments admits a simple circuit, as illustrated by the example in figure 1. This leads to the following decision problem. Given a set of planar line segments S, can one construct a simple circuit from S? This problem has been examined in

[8,9,10] where it has been shown to be NP-complete. In an effort to simplify this question somewhat to permit a polynomial time solution, we propose and solve the following problem.

MONOTONE

Instance: A set of line segments S.

Question: Does S admit a monotone simple circuit?

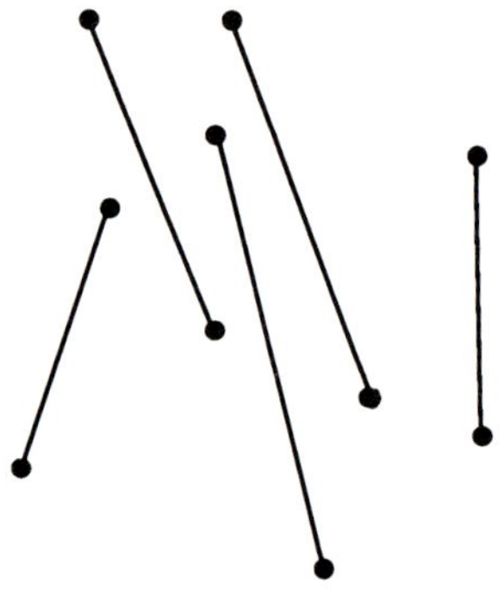

FIGURE 1

To define what is meant by a monotone simple circuit we begin by defining polygonal chains. A *polygonal chain* χ is an alternating sequence of vertices and edges beginning and ending with a vertex, with adjacent edges intersecting at a vertex, and no edges intersecting in their interiors. A polygonal chain χ is said to be *monotone* with respect to a line L if every line orthogonal to L is either collinear to an edge of χ, or intersects χ at most once. A *monotone simple circuit* is a simple circuit that can be decomposed into two polygonal chains beginning and ending at the same vertex, and both chains monotone with respect to the same line. We say that a set of line segments S *admits* a monotone simple circuit if there exists a line L such that a simple circuit that is monotone with respect to L can be constructed from S.

In this paper we show that MONOTONE can be solved in $O(n^4)$ time and $O(n^2)$ space. If a line is pre-specified then we can determine in $O(n^2)$ time and space whether S admits a simple circuit monotone with respect to that given line. We then show that it is sufficient to test for only $O(n^2)$ different lines. The algorithms presented are based on a dynamic programming approach for solving a special case of the travelling salesman problem. This algorithm is used to optimize certain types of Hamiltonian tours in directed graphs. The optimization techniques lead to efficient algorithms ($O(n^4)$ time and $O(n^2)$ space) to optimize the area or the perimeter of the monotone polygon bounded by the monotone simple circuit.

2. PYRAMIDAL TOURS AND MONOTONE CIRCUITS

Define the *polar angle* of a line L to be the angle formed by the x-axis and L. We will be concerned with angles in the range -90° to 90°, where a positive/negative angle is measured counterclockwise/clockwise from the x-axis to L. We will denote a polygon that is monotone with respect to a line with polar angle *d*, a *d*-monotone polygon, and its boundary a *d*-monotone circuit. Fixing *d* to be 0 we specify the following decision problem:

0MONOTONE
INSTANCE: A set of n disjoint line segments S.
QUESTION: Does S admit a 0-monotone circuit?

We prove the following:

Theorem 1: 0MONOTONE can be solved in $O(n^2)$ time and space.

We will also show that:

Theorem 2: If a set of n line segments admits a simple monotone circuit then we can compute the circuit that bounds the minimum (maximum) perimeter 0-monotone polygon, and/or the circuit that bounds the maximum (minimum) area 0-monotone polygon in $O(n^2)$ time and space.

Rather than solve 0MONOTONE directly we propose a similar problem cast in a purely combinatorial setting. We then describe an efficient algorithm for solving this combinatorial problem. Subsequently we show that 0MONOTONE can be transformed into a special case of this problem.

Let G=(V,A), where V= {1,2, ..., n}, be a complete directed graph with an associated cost matrix C, where $C_{i,j}$ represents the cost of the arc (i,j). A *Hamiltonian path* in a graph G is a path that visits every vertex in G exactly once, and can be expressed as a permutation of its vertices. We define a *pyramidal path* as a Hamiltonian path expressible with permutations of the form $(1, i_1, ..., i_r, n, j_1 ..., j_{n-r-2})$, where $i_1 < i_2 ... < i_r < n < j_1 ... < j_{n-r-2}$. A *pyramidal tour* is obtained from a pyramidal path by adding the edge $(j_{n-r-2},1)$. We say a pyramidal tour is optimum if its cost is minimized. Pyramidal tours have been previously described by Gilmore, Lawler and Shmoys in connection with the travelling salesman problem [4]. The sequel closely follows [4]. We formally define the problem:

PYRAMID
INSTANCE: A directed graph G, with an associated cost matrix C.
QUESTION: What is the cost Ω, of an optimum pyramidal tour obtained from G?

A preliminary lemma leads to a dynamic programming algorithm to solve PYRAMID. Let $T_{i,j}$ represent the cost of a shortest pyramidal path that starts at i and ends at j using the vertices {min(i,j), ..., n-1, n}.

Lemma 1:
$$T_{i,j} = \min_{k>j} (C_{i,k} + T_{k,j}) \qquad \text{for } j = i+1, \qquad (1)$$
$$T_{i,j} = \min_{k>i} (C_{k,j} + T_{i,k}) \qquad \text{for } j = i-1, \qquad (2)$$
$$T_{i,j} = C_{i,i+1} + T_{i+1,j} \qquad \text{for } j > i+1, \qquad (3)$$

$$T_{i,j} = C_{j+1,j} + T_{i,j+1} \qquad \text{for } j < i-1. \qquad (4)$$

Proof: To show that (1) holds suppose $\vartheta < T_{i,i+1}$, and there is a pyramidal path from i to i+1 of cost ϑ. Remove the first edge of this pyramidal path, (i,k). Therefore $\vartheta = C_{i,k} + T_{k,i+1}$. Since $T_{i,i+1} = \min_{k>i+1} (C_{i,k} + T_{k,i+1})$, then $\vartheta \geq T_{i,i+1}$ which contradicts $\vartheta < T_{i,i+1}$. To show (3) holds, observe that the edge (i,i+1) is in every pyramidal tour from i to j, where j > i+1. The proofs for equations (2) and (4) mirror those of (1) and (3). □

This lemma leads to the following algorithm:

ALGORITHM PYRAMID

INPUT: A directed graph G=(V,A) with an associated cost matrix C.

OUTPUT: Ω, the value of an optimum pyramidal tour. A table T is constructed, where $T_{i,j}$ represents the value of the minimum cost pyramidal path from i to j. T can later be used to obtain a circuit that realizes Ω.

```
Tn,n-1 ← Cn,n-1; Tn-1,n ← Cn-1,n;
for i ← n-2 downto 1 do
        Ti,i+1 ← min (Ci,k + Tk,i+1); Ti+1,i ← min (Ck,i + Ti+1,k);
                  k>i                           k>i
        for j ← i+2 to n do
                Ti,j ← Ci,i+1 + Ti+1,j; Tj,i ← Ci+1,i + Tj,i+1
        od;
od;
Ω ← min ( (C1,2 + T2,1), (C2,1 + T1,2) );
```

The correctness of algorithm PYRAMID follows from Lemma 1. It is straightforward to show that PYRAMID runs in $O(n^2)$ time and space on an input of a graph with n vertices. To obtain a permutation of the vertices that realizes the optimum pyramidal tour the $O(n^2)$ algorithm described below can be used.

ALGORITHM OBTAIN-TOUR

INPUT: T and Ω as computed by the algorithm PYRAMID.

OUTPUT: A permutation $\Pi = 1, \pi_1, \pi_2, ..., \pi_{n-1}$ corresponding to the optimum pyramidal tour.

```
if Ω = C1,2 + T2,1 then
    π1 ← 2; πn-1 ← k, such that Ck,1 + T2,k = T2,1;
else if Ω = C2,1 + T1,2 then
    πn-1 ← 2; π1 ← k, such that C1,k + Tk,2 = T1,2;
for i ← 2 to n-2 do
    πi ← k, such that Cπi-1,k + Tk,πn-1 = Tπi-1,πn-1;
```

Theorem 3: PYRAMID can be solved in $O(n^2)$ time and space.

Let us examine the problem PYRAMID using instances that are derived from geometric data. Consider a set of n points in the plane sorted on their x-coordinates, and labelled from 1 to n so that $1(x) < 2(x) < ... < n(x)$. We assume that no two points have the same x-coordinate. Handling of the general case is postponed. Let d(i,j) denote the Euclidean distance from the point i to j. We can construct a cost matrix by assigning values $C_{i,j} \leftarrow C_{j,i} \leftarrow d(i,j)$. Solving PYRAMID on this type of input results in a 0-monotone circuit, as is proved by the following lemma.

Lemma 2: Given a cost matrix C, where $C_{i,j} \leftarrow C_{j,i} \leftarrow d(i,j)$, there is a simple 0-monotone circuit that realizes the value Ω obtained by applying algorithm PYRAMID to C.

Proof: Assume we have a pyramidal tour with at least one crossing. It will be shown that we can remove this crossing and remain with a pyramidal tour of no greater length. Assume the edges (a,c) and (b,d) cross as in Figure 2. We can replace (a,c) and (b,d) by the edges (a,b) and (c,d). It is easy to show that the sum of the lengths of the diagonals of a convex quadrilateral are equal to or exceed the sum of the lengths of either pair of opposite sides of the quadrilateral. Replacing these diagonals by the appropriate pair of opposite sides maintains a continuous monotone circuit with length less than or equal to the original. □

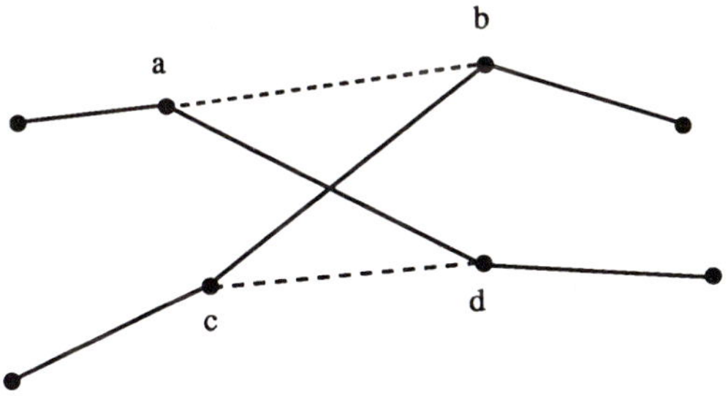

FIGURE 2

If we take the endpoints of a set of segments S, and construct a cost matrix as above, we can no longer guarantee that every optimum pyramidal tour contains all of the required segments, or that the resulting tour is a simple circuit. However, we will be able to enforce these conditions by judiciously assigning values to the cost matrix. In the sequel we assume that our input S is a set of disjoint line segments. It is realistic to forbid sets of segments where two or more segments intersect in their interiors. However, we could permit a set of segments that are openly disjoint, that is, we could allow segments to intersect at their endpoints. The algorithms presented below can be easily modified to allow openly disjoint line segments.

Every monotone circuit can be partitioned into two disjoint polygonal chains each monotone with respect to the same line. In the case of 0-monotone circuits, we call these chains the upper chain and the lower chain for obvious reasons. We will show that there is a property of edges on the upper (lower) chain of every monotone simple circuit that is fundamentally a local property. Given a set of points we can predicate which edges between pairs of points can participate in an upper (lower) chain. We call such edges *safe* upper (lower) edges. As an additional preliminary definition, let α(i,j) denote the polar angle formed by the line through the points i and j. We can can now state the key lemma.

Lemma 3: Let P = (1, 2, ..., n) denote a set of points with 1(x) < 2(x) , ..., < n(x). Every edge, (i,j), j > i, on the upper chain of every 0-monotone simple circuit from P has the property that 1) j = i +1, or 2) α(i,j) > α(i,k) for i < k < j.

Proof: Let the *upper beam* of the segment (i,j), with i(x) < j(x), be the region above (i,j) and bounded on the left and right respectively by the lines y=i(x) and y=j(x). There is a point k ∈ P in the upper beam of (i,j) if and only if α(i,j) < α(i,k) for i < k < j. Assume we have a monotone simple circuit using the edge (i,j) on the upper chain, and the upper beam of (i,j) contains a point k ∈ P. Consider a line L orthogonal to the x-axis through k. This line must also intersect (i,j) and it must also intersect the lower chain of the simple circuit. This implies that L intersects the 0-monotone simple circuit at least three times which contradicts the definition of a monotone simple circuit. Therefore, for every 0-monotone circuit every edge on the upper chain has an empty upper beam thus proving the lemma. □

Clearly safe upper edges must satisfy properties 1) and 2) of the previous lemma. Obviously we can prove a similar result for safe lower edges. That is, every edge, (i,j), j > i, on the lower chain of every 0-monotone simple circuit from P has the property that 1) j = i +1, or 2) α(i,j) < α(i,k) for i < k < j. We can now propose an algorithm that accepts a set of points P and determines a set of safe upper edges U and a set of safe lower edges L.

ALGORITHM SAFE-EDGES
INPUT: A set of points P = (1, 2, ..., n) sorted on their x-coordinates.
OUTPUT: A set of edges U and a set of edges L. The edges in U can safely participate in an upper chain of a monotone circuit and the edges in U can safely participate in a lower chain of a monotone circuit.

```
U ← L ← {(n-1,n)};
for i ← n-2 downto 1 do
    MIN ← MAX ← α(i,i+1); U ← U ∪ {(i,i+1)}; L ← L ∪ {(i,i+1)};
    for j ← i+2 to n do
            if α(i,j) > MAX then U ← U ∪ {(i,j)}; MAX ← α(i,j); fi
            if α(i,j) < MIN then L ← L ∪ {(i,j)}; MIN ← α(i,j); fi
    od;
od;
```

It is not hard to show that algorithm SAFE will construct the sets U and L in $O(n^2)$ time and space. These sets can now assist in the process of assigning weights to edges for algorithm PYRAMID. Let us return to problem 0MONOTONE. Recall we have as input a set of n line segments S. Essentially we would like to find a pyramidal tour through the endpoints of S using all the required edges, that is, the segments S, and create no crossings. We claim that this is a special case of the general pyramidal tour problem. It is a matter of creating the correct cost matrix from the segments S. Assuming $i < j$, let us follow the convention that the cost $C_{i,j}$ represents the cost of the edge (i,j) if it appears on the upper chain of a 0-monotone circuit, and the cost $C_{j,i}$ is the cost if the edge appears on the lower chain of a 0-monotone circuit.

We will assign weights to edges according to their desirability in our pyramidal tour. We can distinguish the edges as defined by all pairs of points i,j into three distinct class, the edges that are required, those that are permissible and those that are forbidden. The required edges are clearly the most desirable and will be assigned an edge weight of 0. An edge that is permissible is one that is in $L \cup U$. To these edges we assign an edge weight of 1. Finally the forbidden edges get an edge weight of 2. Once we have assigned values to all entries in the cost matrix we apply algorithm PYRAMID. If the optimum pyramidal tour is greater than n, then we have either been forced to use a forbidden edge, or an edge that is required has not been used and more than n permissible edges have been used. If the optimum pyramidal tour has a cost of exactly n, then we have used n required edges and n permissible edges. This implies that S admits a 0-monotone circuit. The cost matrix is created in $O(n^2)$ time. The cost of performing algorithm PYRAMID is also $O(n^2)$, so we have solved problem 0MONOTONE in $O(n^2)$ time.

We can actually do more than this, with no increase in asymptotic complexity. If S admits a 0-monotone circuit, then we can compute the cost of a minimum (maximum) perimeter 0-monotone circuit, or the cost of a 0-monotone circuit that bounds the polygon of maximum (minimum) area. Again this is done using algorithm PYRAMID, and simply requires assigning appropriate costs to the cost matrix C. In the case of the minimum perimeter monotone circuit we assign costs as follows. For $1 \leq i < j \leq n$,

$C_{i,j} = 0$	if $(i,j) \in U$ and is required,
$C_{i,j} = d(i,j) + M$	if $(i,j) \in U$ and is not required,
$C_{i,j} = (n+1)M+1$	if $(i,j) \notin U$,
$C_{j,i} = 0$	if $(j,i) \in L$ and is required,
$C_{j,i} = d(i,j) + M$	if $(j,i) \in L$ and is not required,
$C_{j,i} = (n+1)M+1$	if $(j,i) \notin L$.

where $M = \sum_{\forall(i,j)} d(i,j)$. If the result of Ω is less than $(n+1)M$, then no forbidden edges have been used and no more than n permitted edges have been used. The actual cost of the minimum perimeter monotone circuit is $\Omega - nM + \sum_{(i,j)\in S} d(i,j)$.

We can also use this method to compute the 0-monotone simple circuit that bounds the 0-monotone polygon of maximum area. The *convex hull* CH(P) of a set of points P is the smallest area polygon that contains P in its interior. Define the *deficiency polygons* , $D_1, D_2, ..., D_k$, of P as the polygons obtained by subtracting P from CH(P). We can define the area of P as area(P) =

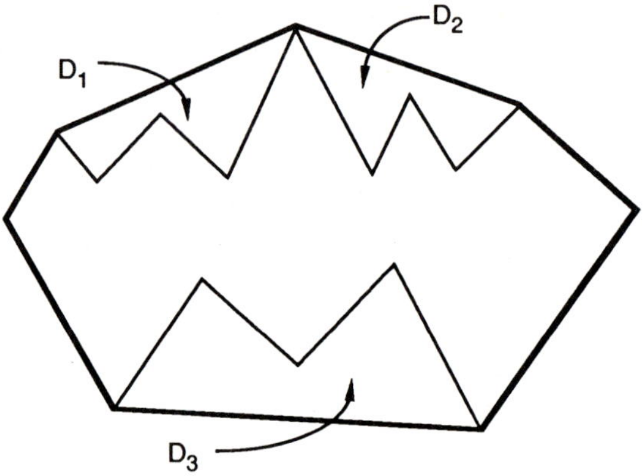

FIGURE 3

area(CH(P)) - ∑area(D_i). See figure 3. Now consider an edge (i,j) in U. Since there are no points in P in the upper beam of (i,j), there are no points in the intersection of CH(P) and the upper beam of (i,j). The area of this region can be easily calculated. Similarly, we can calculate the area of the intersection of the lower beam of an edge in L and CH(P). To compute the 0-monotone simple circuit that bounds the maximum area monotone polygon we assign costs to C as follows:

$C_{i,j} = 0$	if $(i,j) \in U$ and is required,
$C_{i,j}$ = area(upper beam $(i,j) \cap$ CH(P)) + area(CH(P))	if $(i,j) \in U$ and is not required,
$C_{i,j}$ = (n+1)area(CH(P))	if $(i,j) \notin U$,
$C_{j,i} = 0$	if $(j,i) \in L$ and is required,
$C_{j,i}$ = area(lower beam $(i,j) \cap$ CH(P)) + area(CH(P))	if $(j,i) \in L$ and is not required,
$C_{j,i}$ = (n+1)area(CH(P))	if $(j,i) \notin L$.

If the result of Ω is less than (n+1)CH(P), then no forbidden edges have been used and no more than n permitted edges have been used. The actual area of the maximum area monotone polygon is: n+1(CH(P)) - Ω.

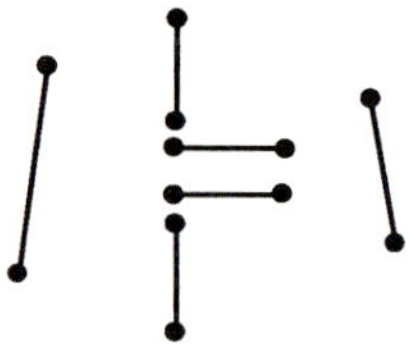

FIGURE 4

We have postponed the issue of handling inputs with two or more endpoints with the same x-coordinate. One cannot simply rotate the input slightly to remove this condition since the original input may admit a 0-monotone circuit and the rotated input may not. In figure 4 a set of segments is shown that admits a 0-monotone circuit as long as it is not rotated. Rotating these segments by any amount (except integer multiples of 180°) precludes a 0-monotone circuit.

Handling this degeneracy is not difficult. However, it results in a rather messy case analysis. Six representative cases are illustrated in figure 5 (a) - (f). Cases (a) and (b) obviate a 0-monotone circuit. Should these cases arise 0MONOTONE is solved immediately. In case (c) the ordering of the endpoints with shared x-coordinates is implicit. Case (d) illustrates the most interesting case. In general, suppose we have k vertical segments in a line denoted by $y=\sigma(x)$. Let the k line segments have their endpoints numbered from top to bottom as $\sigma_1, ..., \sigma_{2k}$. Let $T'\sigma_i,\sigma_j$ represent the cost of a pyramidal tour using all segments on and to the right of the line $y=\sigma(x)$, beginning at σ_i and ending at σ_j. Observe that the only relevant calculations we have to make are $T'\sigma_{2i+1},\sigma_1$, $T'\sigma_{2i+1},\sigma_{2i}$, $T'\sigma_{2k},\sigma_{2i}$, $T'\sigma_{2k},\sigma_1$, for i = 1, ..., k-1 and $T'\sigma_{1,j}$, $T'\sigma_{2k,j}$, T'_{j},σ_1, T'_{j},σ_{2K}, for all points $j \in S$ such that $j(x) > \sigma(x)$. The cost of computing these values is O(kn) time using straightforward methods. This maintains the overall complexity of the previous algorithms at $O(n^2)$ time and space. The cases in figures 5(e) and 5(f) are handled in much the same way as case 5(d). The details are omitted.

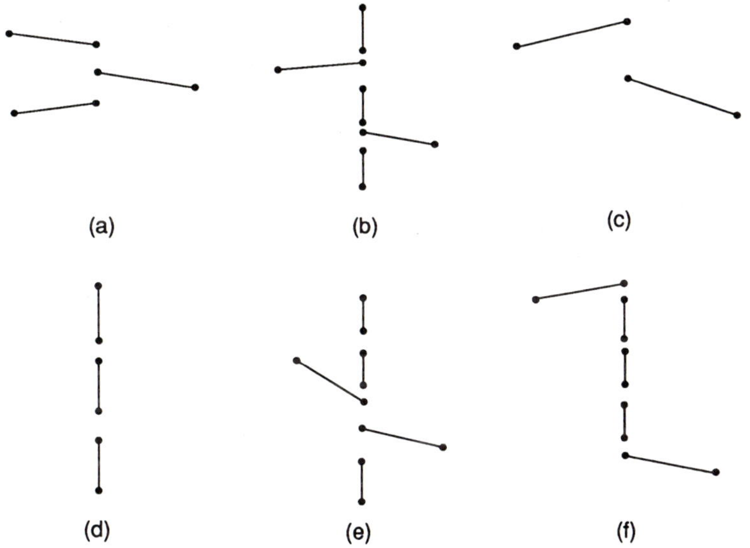

FIGURE 5

We return to problem MONOTONE. Given a set of n segments, S, we would like to know if there exists any line of direction d such that S admits a d-monotone simple circuit. A line with polar angle d will be denoted L_d. The following observation is a prelude to the key lemma.

Observation : Given a line segment with endpoints s,t (assume s is above t) consider the orthogonal projections of s and t on the lines L_a, L_b and L_c, where L_b is normal to the segment (s,t), and the angles a,b,c have the following relationship: $a < b < c$. The orthogonal projections of s and t on these lines have the property that on L_a, s precedes t, on L_b, s and t have the same coordinates, and on L_c, t precedes s.

This observation implies that the line normal to a segment is critical in determining a change in order. This fact will be used to prove the next lemma.

Lemma 4: Given a polygon P and a set of lines L^*, where the normal to every edge in P can be found in L^*, if P is a monotone polygon, then P must be monotone with respect to a line in L^*.

Proof: We contradict the assumption that P is monotone with respect to a line L_δ, and L_δ is not in L^*. Consider the line L_d where $L_d \in L^*$, and d is the smallest angle greater than δ. We can assume without loss of generality that the non-monotonicity of P with respect to L_d interchanges the vertices s and t, an edge of P, in the orthogonal projection on L_d. But this is impossible! As was observed above, the order of orthogonal projections of endpoints of a segment (s,t) differ on a line L_a from a line L_c only if the line normal to (s,t), L_b, lies between L_a and L_c. Since L_d is the closest normal line (from above) to L_δ, the normal to (s,t) cannot be between these two lines. If P is monotone with respect to L_δ, then it is also monotone with respect to L_d. □

The previous lemma leads to an algorithm to determine if a set of segments S admits a d-monotone simple circuit for arbitrary d. Consider fixing a point o in the interior of the convex hull of S. Now we can rotate the segments about this point o so that after rotating re-application of 0MONOTONE will test for monotonicity for a direction different from before. We simply rotate the points by the appropriate amounts so that each line in L(S) becomes the new x-axis. Therefore, by solving 0MONOTONE on $O(n^2)$ rotations of S is sufficient to determine whether S admits a d-monotone polygon for arbitrary d. The results that apply to attaining optimal 0-monotone polygons with respect to perimeter and area apply here as well. As a result we can state the following theorem.

Theorem 3: We can determine whether a set of segments S admits a monotone simple circuit in $O(n^4)$ time and $O(n^2)$ space. We can also obtain a monotone polygon which optimizes area or perimeter using the same amount of time and space.

3. DISCUSSION

As was mentioned in the introduction it is NP-complete to determine whether a set of line segments admits a simple circuit. We have solved a restricted case of this problem in polynomial time. The initial impetus for looking at the monotone case was inspired by the following problem. Given a set of vertical line segments determine if they admit a simple circuit. Using only vertical line segments seems very limiting. In fact after some thought the following conjecture comes to mind.

Conjecture : A set of vertical line segments S admits a simple circuit if and only if S admits a 0-monotone simple circuit.

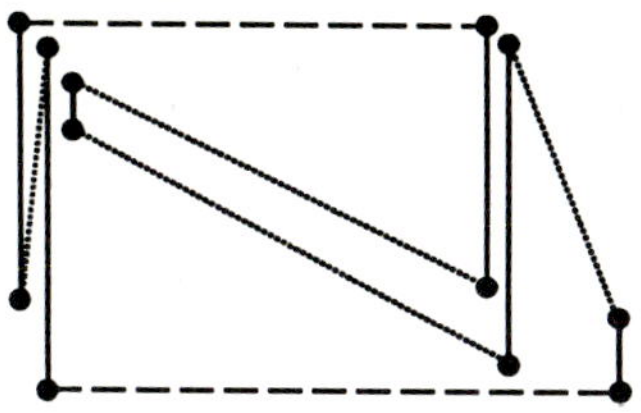

FIGURE 6

The if part is trivially true. It remains to prove that if S admits a simple circuit, then S also admits a 0-monotone simple circuit. However, this conjecture is false! The counter example is shown in figure 6. In figure 6, hatched edges represent connections that are forced, and the dotted edges complete a non-monotone simple circuit. It is unknown whether deciding if a set of disjoint vertical line segments admits a simple polygon is NP-complete or not.

Another open problem is to compute the minimum number of disjoint monotone simple circuits obtainable from a set of line segments. In [6] it is suggested one could obtain some perceptual information from the unique set of orthogonal polygons obtained from a set of points. In a similar vein a minimum set of monotone simple circuits from a set of line segments may provide some perceptually useful information.

Finally it may be possible to improve the complexity of the algorithm presented here to solve the MONOTONE problem. We repeat an $O(n^2)$ algorithm $O(n^2)$ times making no use of previously obtained information in subsequent iterations. Perhaps one could use some of this previously gained information to improve on the running time of the algorithm on the whole.

REFERENCES

[1] H. ElGindy, G.T. Toussaint, "On triangulations of line segments," McGill University Technical Report No. SOCS-84.9, 1984.

[2] I. Fáry, "On straight-line representations of planar graphs," *Acta Sci. Math. Szeged.* **11** (1948) 229-233.

[3] M. R. Garey, D. S. Johnson, R. E. Tarjan, "The planar Hamiltonian circuit problem is NP-complete," *SIAM J. Computing* **5** (1976) 704-714.

[4] P.C. Gilmore, E.L. Lawler, D.B. Shmoys, "Well solved special cases," in *The Travelling Salesman Problem,* edited by E. L. Lawler, J. K. Lenstra, A. H. G. Rinnooy Kan, D. B. Shmoys, John Wiley and Sons, (1985) 87-143.

[5] A. Itai, C.H. Papadimitriou, J.L. Szwarcfiter, "Hamiltonian paths in grid graphs," *SIAM J. Computing* **11**(4) (1982) 676-686.

[6] J. O'Rourke, "Uniqueness of orthogonal connect-the-dots," this volume.

[7] J. O'Rourke, H. Booth, R. Washington, "Connect-the-dots: a new heuristic," Technical Report JHU/EECS-84/11, Johns Hopkins University (1984).

[8] D. Rappaport, "The Complexity of Computing simple Circuits in the Plane," Ph.D. Dissertation, McGill University, (December 1986).

[9] D. Rappaport, "Computing simple circuits from a set of line segments is NP-complete," *the 3rd ACM Symposium on Computational Geometry,* Waterloo, Ontario (June 1987) 52-60.

[10] D. Rappaport, H. Imai, G. T. Toussaint, "On computing simple circuits on a set of segments," *the 2nd ACM Symposium on Computational Geometry,* Yorktown Heights, New York (June 1986) 322-330.

[11] G. Toussaint, "Pattern recognition and geometrical complexity," in *Proc. 5th International Conference on Pattern Recognition,* Miami Beach (December 1980) 1324-1347.

[12] G. T. Toussaint, "Computational Geometry and Morphology," Technical Report TR-SOCS-86.3, McGill University (February 1986).

Computational Morphology
G.T. Toussaint (editor)
Elsevier Science Publishers B.V. (North-Holland), 1988

Circular Separability of Planar Point Sets

Binay K. Bhattacharya+

ABSTRACT

Two planar point sets S_1 and S_2 are circularly separable if there is a circle that encloses S_1 but excludes S_2. In this note an $O((|S_1|+|S_2|)\log(|S_1|+|S_2|))$ algorithm is presented to determine the set $S(S_1,S_2)$ of points which are centers of circles that enclose S_1 but exclude S_2. It is then shown that once $S(S_1,S_2)$ is known, all the smallest and the largest separating circles can be computed very easily.

1.Introduction

Let S_1 and S_2 be two planar sets of points. We are interested in computing the set $S(S_1,S_2)$ defined as follows:

$$S(S_1,S_2) = \{ x \mid \max_{\forall p \in S_1} d(x,p) \le \min_{\forall q \in S_2} d(x,q)\}.$$

Thus for every $x \in S(S_1,S_2)$ there exists a circle C_x centered at x such that each data point of S_1 is interior to or on the boundary of C_x while each data point of S_2 is exterior to or on the boundary of C_x. We will refer such a circle as a separating circle which includes S_1 and excludes S_2. If $S(S_1,S_2)$ is not empty, S_1 and S_2 are said to be circularly separable.

The problem of whether S_1 and S_2 are circularly separable has applications in pattern recognition and image processing [KA]. A *digital disk* is defined to be the set of lattice points (points with integer coordinates) which are contained inside or on some circle [KA]. The shape recognition problem is to determine whether a given set D of lattice points is a digital disk. Let S_1 be the set of lattice points representing the boundary of D. Let S_2 be the set of all lattice points exterior to D but 8-connected to a point of S_1, that is, adjacent horizontally, vertically, or diagonally. Now D is a digital disk if and only if there exists a separating circle for S_1 and S_2. Kim and Anderson [KA] presented an $O((|S_1|+|S_2|)^2)$ algorithm to determine a separating circle for S_1 and S_2. O'Rourke and Rao Kosaraju [OR] solved the smallest separating circle problem in $O(|S_1| + |S_2|)$ time by transforming this problem to a convex quadratic minimization problem in three dimensions. Recently, Fisk [F] determined the set $S(S_1,S_2)$ in $O(\max(|S_1||S_2|, (|S_1|+|S_2|)\log(|S_1|+|S_2|))$ time.

+School of Computing Science, Simon Fraser University, Burnaby, B.C., Canada, V5A 1S6.
Research was supported by grant NSERC A2422.

In this note an $O((|S_1| + |S_2|)\log(|S_1| + |S_2|))$ algorithm to determine $S(S_1,S_2)$ is presented. We then show that all the smallest and the largest separating circles can be determined easily once the set $S(S_1,S_2)$ is known.

In section 2 various properties of $S(S_1,S_2)$ are described. Section 3 formally presents the algorithm CIRCULAR to compute $S(S_1,S_2)$, along with the worst-case analysis. Section 4 deals with the problem of computing all the smallest and the largest separating circles for S_1 and S_2.

2. Geometric properties of $S(S_1,S_2)$

We assume that both S_1 and S_2 are non-empty. When S_1 or S_2 is empty, the problem of computing $S(S_1,S_2)$ is not well defined.

Let $VD_{FP}(S_1)$ denote the furthest-point Voronoi diagram of S_1. Let $VD_{NP}(S_2)$ denote the nearest-point Voronoi diagram of S_2. These geometrical structures are well known and the various properties of these structures are discussed in [PS, BT]. We can describe $VD_{FP}(S_1)$ and $VD_{NP}(S_2)$ as follows:

$$VD_{FP}(S_1) = \bigcup_{\forall p \in S_1} W_p(S_1), \text{ and}$$

$$VD_{NP}(S_2) = \bigcup_{\forall q \in S_2} V_q(S_2),$$

where

$$W_p(S_1) = \{x \mid d(x,p) \geq d(x,r) \text{ for all } r \in S_1\}, \text{ and}$$

$$V_q(S_2) = \{x \mid d(x,q) \leq d(x,r) \text{ for all } r \in S_2\}.$$

Fisk [F] has defined $S(S_1,S_2)$ as

$$S(S_1,S_2) = \bigcup_{\substack{\forall p \in S_1 \\ \forall q \in S_2}} [\, W_p(S_1) \cap V_q(S_2) \cap H(p,q)] \qquad (1)$$

where $H(p,q)$ is the half-plane containing p determined by the perpendicular bisector of p and q, $B(p,q)$. Thus $S(S_1,S_2)$ is the set of all points in the plane which are centers of separating circles for S_1 and S_2. From (1) it is clear that the boundary of S(S_1,S_2) is polygonal.

Lemma 1: The set S(S_1,S_2) is convex.

Proof. Clearly, the lemma is true if S(S_1,S_2) is either empty or a set with just one point. Let u and v be two points of S(S_1,S_2). There exist circles C_u and C_v with the centers at u and v, respectively, which include S_1 but exclude S_2 (Fig.1). Clearly, the data points of S_1 must lie in the intersection $C_u \cap C_v$. For any arbitrary point y on **uv**, the circle C_y with the center at y and radius d(y,s) (refer Fig.1) contains $C_u \cap C_v$ and is contained in $C_u \cup C_v$. Therefore, y is the center of a separating circle for S_1 and S_2. Hence, the entire line segment **uv** must belong to S(S_1,S_2). Since this is true for all u and v in S(S_1,S_2), it follows that S(S_1,S_2) is convex. □

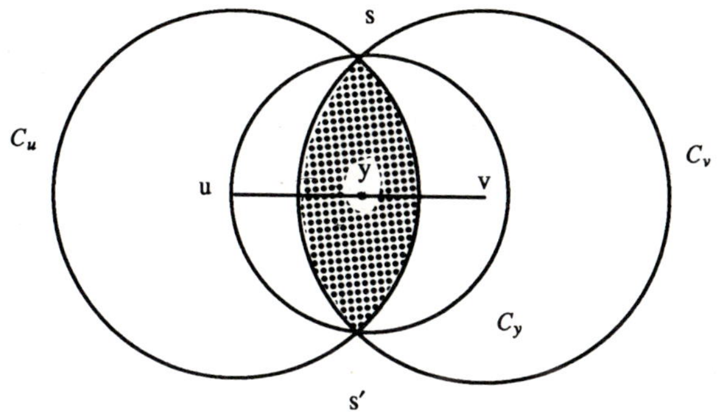

Figure 1

Fig.2 illustrates different forms of S(S_1,S_2).

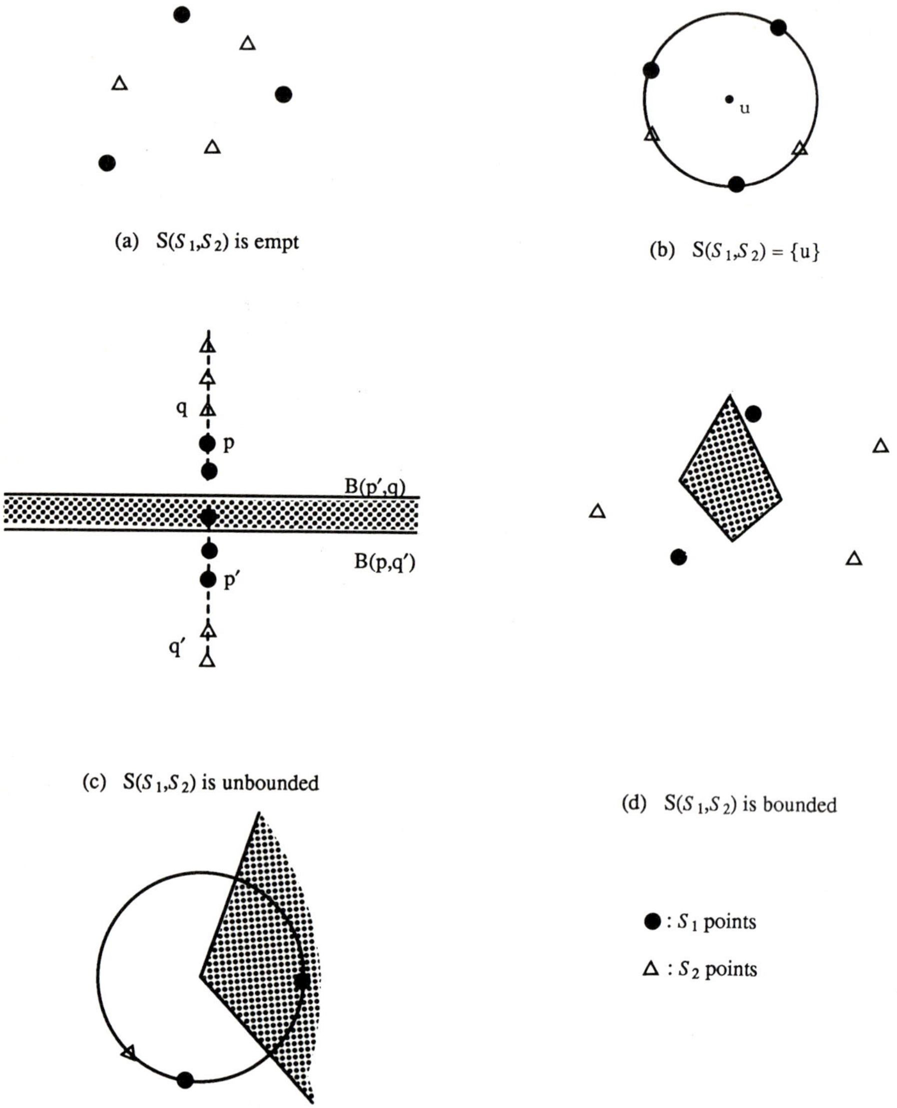

(a) $S(S_1,S_2)$ is empt

(b) $S(S_1,S_2) = \{u\}$

(c) $S(S_1,S_2)$ is unbounded

(d) $S(S_1,S_2)$ is bounded

(e) $S(S_1,S_2)$ is unbounded

Figure 2

Lemma 2: If the points of $S_1 \cup S_2$ don't all lie on a line, S(S_1,S_2), if it exists, contains at least one vertex.

Proof: Without any loss of generality, let p, q and r be three data points determining a nondegenerate triangle with $p, r \in S_1$ and $q \in S_2$. Then by definition (1), S(S_1,S_2) $\subseteq$ H(p,q) $\cap$ H(r,q). Since the boundary of S(S_1,S_2) is polygonal, S(S_1,S_2), if exists, must contain at least one vertex.
□

We now show that S(S_1,S_2) satisfies the following two lemmas. (We assume henceforth that S(S_1,S_2) exists.)

Lemma 3: A point u is an extreme point of S(S_1,S_2) if and only if there exists a separating circle, centered at u, which is the circumcircle of three points of $S_1 \cup S_2$ with at least one point from each set.

Lemma 4: A point u is a boundary point of S(S_1,S_2) if and only if there exists a separating circle, centered at u, which passes through two points of $S_1 \cup S_2$ with one point from each set.

Proof of Lemma 3. Let u be an extreme point of S(S_1,S_2). We can conclude from the definition of S(S_1,S_2), given by the equation (1), that u could be a vertex of $VD_{FP}(S_1)$ or $VD_{NP}(S_2)$ or a point of intersection of B(p,q) with the boundary of either $W_p(S_1)$ or $V_q(S_2)$ for some $p \in S_1$ and $q \in S_2$. However, it is not difficult to see that u can not be a vertex of either $VD_{FP}(S_1)$ or $VD_{NP}(S_2)$. If possible let u be a vertex of $VD_{FP}(S_1)$, not lying on any edge of $VD_{NP}(S_2)$. u is then the circumcenter of some group of at least three points of S_1, and the corresponding circumcircle C_u encloses all the points of S_1. Let q be the point of S_2 which is nearest to u. q must lie outside the circle C_u. Then there exists a neighborhood of u such that for any point x in the neighborhood, x $\in$ S(S_1,S_2) (Fig.3). Hence u is not an extreme point of S(S_1,S_2) - a contradiction. Therefore, u can not be a vertex of $VD_{FP}(S_1)$ only. Similarly, we can show that u can not be a vertex of $VD_{NP}(S_2)$ only. Therefore, any vertex of S(S_1,S_2) is the result of an intersection of B(p,q) with the boundary of $W_p(S_1) \cap V_q(S_2)$, for some $p \in S_1$ and $q \in S_2$.

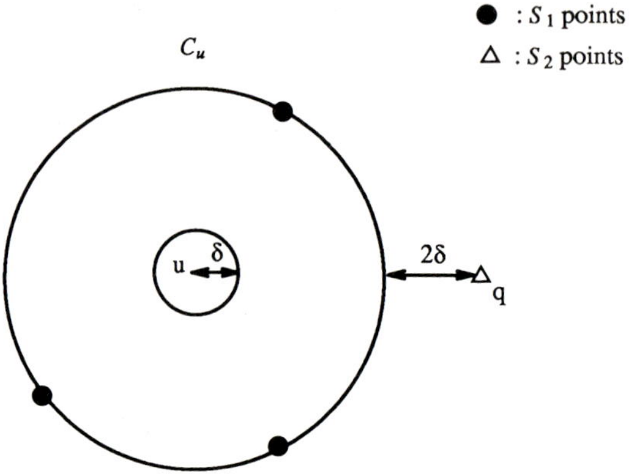

Figure 3

Suppose u is the intersection of $B(p,q)$ and the boundary of $W_p(S_1)$. Then u is the intersection of $B(p,q)$ and $B(p,r)$ for some $r \in S_1$. Therefore, u is the circumcenter of p, q and r where $p,r \in S_1$ and $q \in S_2$ and the corresponding circumcircle is a separating circle for S_1 and S_2. Similarly, when u is the intersection of $B(p,q)$ and the boundary of $V_q(S_2)$, u is the circumcenter of p,q and some point $w \in S_2$ and the corresponding circumcircle contains S_1 but excludes S_2.

Suppose u is the circumcenter of p, q and r where $p \in S_1$, $q \in S_2$ and $r \in S_1 \cup S_2$ and the corresponding circumcircle is a separating circle. According to the definition of $S(S_1,S_2)$, $u \in S(S_1,S_2)$. When $r \in S_1$, $S(S_1,S_2) \subseteq H(p,q) \cap H(r,q)$, and $S(S_1,S_2) \subseteq H(p,q) \cap H(p,r)$ when $r \in S_2$. In each of the two cases there exists a supporting line to $S(S_1,S_2)$ through u. Hence u is an extreme point of $S(S_1,S_2)$. □

Using similar arguments one can easily prove lemma 4.

We now prove the following lemma which establishes the size of $S(S_1,S_2)$.

Lemma 5: The number of vertices of $S(S_1,S_2)$ is $O(|S_1| + |S_2|)$.

Proof: Lemma 3 implies that the vertices of $S(S_1,S_2)$ must lie on the Voronoi edges of $VD_{FP}(S_1)$ and $VD_{NP}(S_2)$. Since $S(S_1,S_2)$ is convex, the number of vertices of $S(S_1,S_2)$ that can lie on any Voronoi edge is at most two. Since there are $O(|S_1| + |S_2|)$ edges in $VD_{FP}(S_1)$ and $VD_{NP}(S_2)$, the number of vertices in $S(S_1,S_2)$ is $O(|S_1| + |S_2|)$. □

3. Algorithm CIRCULAR and its worst-case analysis

We will assume that $|S_1| > 0$ and $|S_2| > 0$. We also assume that all the data points of $S_1 \cup S_2$ do not lie on a straight line. If it is so [see Fig. 2(c)], we can determine $S(S_1,S_2)$, if exists, in $O(|S_1| + |S_2|)$ time.

The step by step description of the algorithm CIRCULAR to compute $S(S_1,S_2)$ is described below.

Input: Sets S_1 and S_2.

Output: The boundary of $S(S_1,S_2)$ as a convex polygon.

Begin

Step 1: Determine a point $x_0 \in S(S_1,S_2)$. If no such point exists then stop. $\{S(S_1,S_2) = \Phi\}$

Step 2: Determine an extreme point $u_0 \in S(S_1,S_2)$.

Step 3: Construct $VD_{FP}(S_1)$ and $VD_{NP}(S_2)$.

Step 4: Set u $\leftarrow u_0$.

Step 5: Determine the vertex w of $S(S_1,S_2)$ which is the counterclockwise neighbor of u.

Step 6: If w $\equiv u_0$, stop.

Step 7: If w is not a point at infinity, set u $\leftarrow$ w and go to step 5.

Step 8: {w is a point at infinity.}

Set u $\leftarrow u_0$

Step 9: Determine the vertex w of $S(S_1,S_2)$ which is the clockwise neighbor of u.

Step 10: If w is not a point at infinity, set u $\leftarrow$ w and go to step 9. Otherwise, stop.

End

3.1 Implementation of CIRCULAR

We assume that $|S_1| = n_1$ and $|S_2| = n_2$. Here we discuss in details the implementations of steps 1, 2 and 5. Step 9 is similar to step 5. Step 3 can be solved in $O(n_1 \log n_1 + n_2 \log n_2)$ by using the algorithms proposed by Shamos [S], Brown [B]. The remaining steps are easy to implement.

3.1.1 Step 1

O'Rourke and Rao Kosaraju [OR] have shown that the problem of finding a separating circle of S_1 and S_2 is equivalent to the problem of finding a separating plane of S'_1 and S'_2 in three dimensions, where S'_i contains points of the form (x, y, x^2+y^2) for all $(x,y) \varepsilon S_i$, i=1,2. It is well known that linear separability problem can be transformed to a linear programming problem [DR]. Using the linear programming approach, one can find a separating plane of S'_1 and S'_2 in $O(n_1+n_2)$ time [M,D]. Therefore, we can determine x_0, if exists, in $O(n_1+n_2)$ time. However this method has a very large constant of proportionality.

There exists a relatively simple nontrivial algorithm to determine the separating plane of S'_1 and S'_2 in $O(n_1 \log n_1 + n_2 \log n_2)$ time. This is achieved by first constructing the convex hulls of S'_1 and S'_2 using the algorithm of Preparata and Hong [PH] as modified by Edelsbrunner [E] and then applying the linear time algorithm of Dobkin and Kirkpatrick [DK] to find a separating plane of two convex polyhedra. Therefore, in this case we can implement Step 1 in $O((n_1+n_2)\log(n_1+n_2))$ time.

3.1.2 Step 2

Once x_0 is known, we can determine the separating circle, with the center at x_0, which passes through a point $q \varepsilon S_2$ in $O(n_2)$ time. If this circle also passes through a point of S_1 and another point of $S_1 \cup S_2$, then according to lemma 3, x_0 is a vertex of $S(S_1,S_2)$. In this case $u_0 \equiv x_0$. Otherwise, assume that the separating circle with the center at x_0 passes through only $q \varepsilon S_2$. We can then find another point y on the line segment x_0q such that the circle with the center at y and radius $d(y,q)$ is a separating circle which also passes through a point $p \varepsilon S_1$. According to lemma 4, y is a boundary point of $S(S_1,S_2)$. y can easily be determined in $O(n_1)$ time. The method is similar to the one described in [AB].

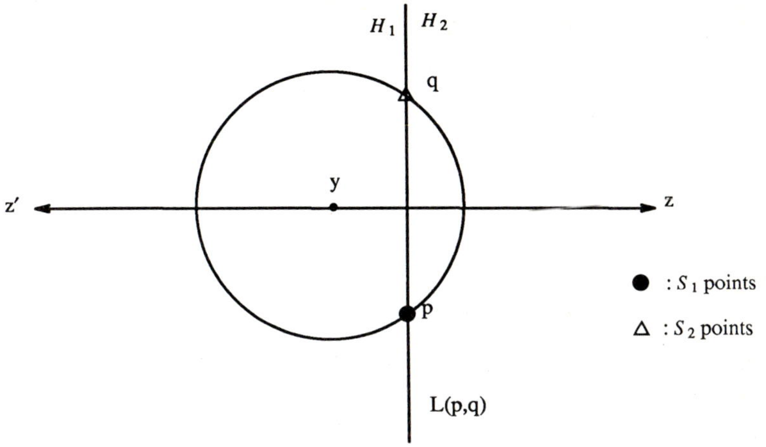

Figure 4

Suppose L(p,q) be the line through p and q which determines two open half-planes H_1 and H_2 where H_1 contains y (Fig.4). If $(S_1 \cap H_1) \cup (S_2 \cap H_2)$ is empty, for any point w on the ray **yz**, the circle with the center at w and radius d(w,p) is a separating circle for S_1 and S_2. Hence **yz** is contained by some unbounded edge of S(S_1,S_2). Similarly, if $(S_2 \cap H_1) \cup (S_1 \cap H_2)$ is empty, the ray **yz′** is contained by some unbounded edge of S(S_1,S_2). Since the points of $S_1 \cup S_2$ don't lie on a straight line, without any loss of generality, let $(S_1 \cap H_1) \cup (S_2 \cap H_2)$ be non empty. We can then find the circle in O(n_1+n_2) time which passes through p and q and another distinct point of $(S_1 \cap H_1) \cup (S_2 \cap H_2)$ such that its center w is closest to y. Again the method of determining such circle is similar to the one described in [AB]. Clearly, w lies on **yz** and the circle is a separating circle which passes through p, q and a point of $S_1 \cup S_2$. Therefore, w is a vertex of S(S_1,S_2) and w can be determined in O(n_1+n_2) time. Therefore, step 2 can be implemented in O(n_1+n_2) time.

3.1.3 Step 5

We will initially assume here that the points of S_1 and S_2 are in general positions i.e. no four points of $S_1 \cup S_2$ are cocircular. Later we will relax this constraint.

We know that the boundary of $S(S_1,S_2)$ is either a bounded or unbounded convex polygon. First we are interested in determining the edge of $S(S_1,S_2)$ incident on the vertex u which contains the counterclockwise neighbor of u. We already know that, for some $p \in S_1$ and $q \in S_2$,

(i) $\{u\} = B(p,q) \cap bd(W_p(S_1))$, or

(ii) $\{u\} = B(p,q) \cap bd(V_q(S_2))$.

Case 1: $\{u\} = B(p,q) \cap bd(W_p(S_1))$

In this case u is the circumcenter of p, q and some point $r \in S_1$. Therefore, u is an interior point of $V_q(S_2)$. Let **uv** be the ray equidistant to p and q such that the circle with center y on **uv** and radius $d(y,p)$ includes the point r (Fig.5). It is now possible to find a point z on **uv** at a distance δ from u, for an arbitrarily small δ, such that the circle with the center at z and radius $d(z,p)$ is a separating circle. Therefore, according to lemma 4, z is a boundary point. Hence **uv** contains an edge of $S(S_1,S_2)$ incident on u. Similarly, we can show that **uv′** contains another edge of $S(S_1,S_2)$ incident on u (Fig.5). We take the edge that contains the counterclockwise neighbor of u. From Fig.5 we see that the ray **uv** contains the counterclockwise neighbor of u.

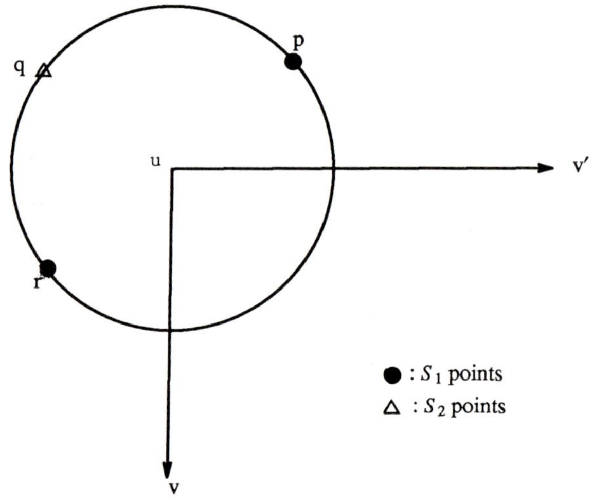

Figure 5

We know that u is an interior point of $V_q(S_2)$. We also know that u lies on the boundary of $W_p(S_1)$ and $W_r(S_1)$. Therefore the ray **uv** will first intersect the interior of $W_p(S_1)$. Let x_{FP} be the other intersection point of **uv**, if exists, with bd($W_p(S_1)$). If there is no such intersection, we consider x_{FP} to be a point at infinity. Let x_{NP} be the intersection point of **uv** with bd($V_q(S_2)$). Again if there is no such intersection, we consider x_{NP} to be a point at infinity. We can report x_{FP} and x_{NP} in O($\log n_1$) and O($\log n_2$) time respectively by first preprocessing each Voronoi polygon of $VD_{FP}(S_1)$ and $VD_{NP}(S_2)$ [AES, C]. The preprocessing step requires O(n_1+n_2) time once $VD_{FP}(S_1)$ and $VD_{NP}(S_2)$ are known. Let $x_{\min}$ be the intersection point nearest to u. We then have the following result.

Lemma 6: $x_{\min}$ is an extreme point of S(S_1,S_2) and is the counterclockwise neighbor of u.

Proof. We first assume that $x_{\min}$ is not a point at infinity. If $x_{\min} \equiv x_{FP}$, $x_{\min}$ is an interior point of $VD_q(S_2)$ and $x_{\min}$ lies on a boundary edge of $V_{FP}(S_1)$ which is determined by the perpendicular bisector B(p,t) for some $t \in S_1$. Thus $x_{\min}$ is the circumcenter of p, q and t and the corresponding circumcircle encloses S_1 but excludes S_2. Hence, according to lemma 3, $x_{\min}$ is an extreme point of S(S_1,S_2). Similarly, if $x_{\min} \equiv x_{NP}$, x is the circumcenter of p, q and a point $t \in S_2$. Also $x_{\min}$ is an interior point of $W_p(S_1)$. Hence the circumcircle of p, q and t encloses S_1 and excludes S_2. In this case $x_{\min}$ is an extreme point of S(S_1,S_2).

Let us now assume that $x_{\min}$ is a point at infinity. Since the unbounded line **uv** lies completely in the interior of $W_p(S_1)$ and $V_q(S_2)$, for any point y on **uv**, the circle with the center at y and radius d(y,p) separates S_1 and S_2. Hence, according to lemma 3, y is a boundary point of S(S_1,S_2). Hence **uv** is an unbounded edge of S(S_1,S_2). □

Case 2: $\{u\} = B(p,q) \cap \text{bd}(V_q(S_2))$

In this case u is the circumcenter of p, q and some point $t \in S_2$(Fig.6). As before, we can show that the edges of S(S_1,S_2) incident on u are contained in **uv** and **uv′** (Fig.6). The counterclockwise neighbor of u can be determined in similar fashion as described in Case 1.

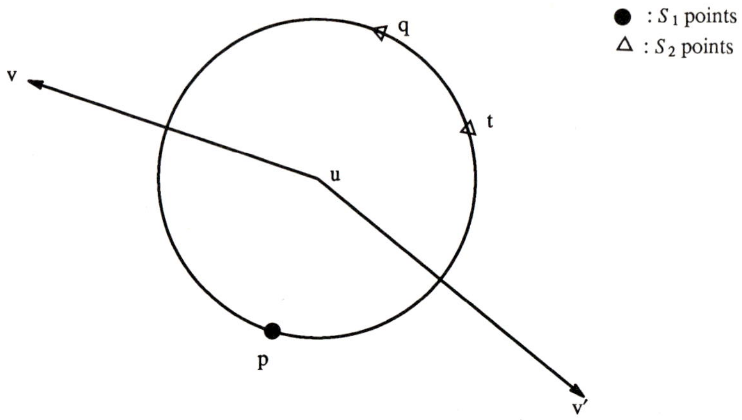

Figure 6

We now relax the constraint that no more than three points could be cocircular. Let u be an extreme point of $S(S_1,S_2)$. Let X_1 and X_2 be the set of points of S_1 and S_2, respectively, lying on the separating circle with the center at u. Let $|X_1| = t_1$ and $|X_2| = t_2$. Clearly, $t_1 > 0$ and $t_2 > 0$, otherwise, u can not be an extreme point of $S(S_1,S_2)$. If $t_1 > 2$, u is a vertex of $VD_{FP}(S_1)$. In this case X_1 can be determined in $O(t_1)$ time from $VD_{FP}(S_1)$. Similarly, X_2 can be determined in $O(t_2)$ time. Let $CH(X_i)$ denote the convex hull of the set X_i, i = 1,2. If $CH(X_1)$ and $CH(X_2)$ do not intersect, the edges of $S(S_1,S_2)$ incident on u will be contained by **uv** and **uv′** as shown in Fig.7. Since all the points of X_1 and X_2 lie on a circle, we can determine in $O(t_1+t_2)$ time whether $CH(X_1)$ and $CH(X_2)$ intersect. If $CH(X_1)$ and $CH(X_2)$ intersect, it is not difficult to see that $S(S_1,S_2) = \{u\}$.

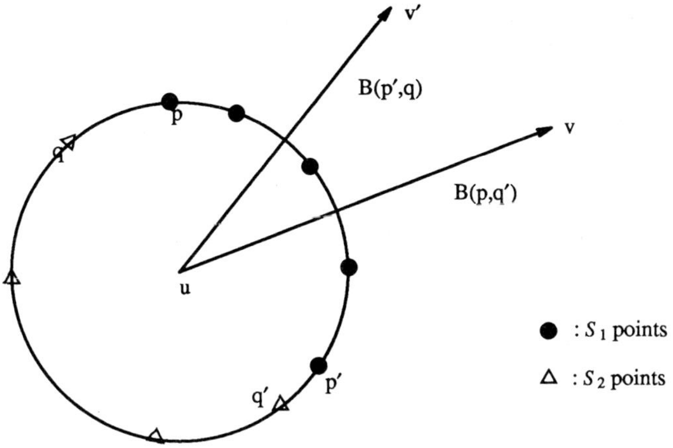

Figure 7

The preprocessing of the Voronoi polygons will be done exactly once. Since the size of $S(S_1,S_2)$ is $O(n_1+n_2)$, Step 5 will be executed at most $O(n_1+n_2)$ times. Therefore, the worst-case running time to execute Step 5 is $O((n_1+n_2)\log(n_1+n_2))$.

3.2 Worst-case running time of CIRCULAR

We have discussed in section 3.1 an implementation of the steps of CIRCULAR. It is clear fronm the description that that the worst-case running time of CIRCULAR is $O((n_1+n_2)\log(n_1+n_2))$ and the storage requirement is $O(n_1+n_2)$.

4. Smallest and largest separating circles.

O'Rourke and Rao Kosaraju [OR] gave $O((n_1+n_2)\log(n_1+n_2))$ algorithms to compute all the smallest and the largest separating circles for S_1 and S_2. These algorithms are quite complicated. In this section we show that once $VD_{FP}(S_1)$, $VD_{NP}(S_2)$ and $S(S_1,S_2)$ are known, the smallest and the largest separating circles can be determined very easily.

The following two lemmas form the basis of our algorithm.

Lemma 7: A smallest separating circle for S_1 and S_2 is either the minimum spanning circle of S_1 or it must pass through two points of S_1 and one point of S_2.

Lemma 8: [OR] A largest separating circle for S_1 and S_2 must pass through three points of S_2 or it must pass through two points of S_2 and one point of S_1.

Proof of Lemma 7

Clearly, if the center of the minimum spanning circle (MSC) of S_1 belongs to $S(S_1,S_2)$, this circle is the smallest separating circle for S_1 and S_2. There will be only one such circle since the minimum spanning circle of a set is unique.

Suppose the center of the MSC does not belong to $S(S_1,S_2)$. Let w be the center of a smallest separating circle C_w. If C_w passes through exactly one point of S_1 or two points of S_1 (two points are not diametrically opposite) or three points of S_1 (three points form an obtuse angle triangle) or one point of S_2 or two points of S_2 or three points of S_2, then we can construct a smaller separating circle from C_w. Again if C_w passes exactly one point of S_1 and one or more points of S_2, we can similarly construct a smaller separating circle from C_w. Therefore, C_w must pass through two points of S_1 and one point of S_2. □

Clearly, if $S(S_1,S_2)$ is unbounded, the largest separating circle has radius infinity. We assume that $S(S_1,S_2)$ is bounded. According to lemma 4, each vertex of $S(S_1,S_2)$ is a point equidistant to three points of $S_1 \cup S_2$ with at least one point from each set. Therefore, lemmas 7 and 8 imply that the centers of smallest and largest separating circles for S_1 and S_2 can be obtained by only examining the vertices of $S(S_1,S_2)$, the minimum spanning circle of S_1 and the vertices of $VD_{NP}(S_2)$ which are inside $S(S_1,S_2)$. We can determine the minimum spanning circle of S_1 in $O(n_1)$ time from $VD_{FP}(S_1)$ [BT]. We can determine the vertices of $VD_{NP}(S_2)$ inside $S(S_1,S_2)$ in $O((n_1+n_2)\log(n_1+n_2))$ time [S]. Therefore, all the smallest and the largest separating circles can be determined in $O((n_1+n_2)\log(n_1+n_2))$ time with linear storage requirement.

5. Conclusions

We presented a simple $O((|S_1| + |S_2|)\log(|S_1 + |S_2|))$ algorithm to compute the set $S(S_1,S_2)$. We then showed that once $S(S_1,S_2)$ is known, all the smallest and the largest separating circles for S_1 and S_2 can be determined very easily. The algorithm makes an efficient search in both the furthest-point and nearest-point Voronoi diagrams. This approach does not naturally extend to higher dimensions. Efficient computation of $S(S_1,S_2)$ in higher dimension is still an open problem.

REFERENCES

[AB] Avis, D. and Bhattacharya, B.K., "Algorithms for computing d-dimensional Voronoi diagrams and their duals," *Advances in Computing Research*, Vol. 1, Edited by Preparata, F.P., pp.159-180, 1983.

[AES] Avis, D., ElGindy, H. and Seidel, R., "Simple online algorithms for convex polygons," *Computational Geometry*, Edited by Toussaint, G.T., North-Holland, pp.23-42, 1985.

[BT] Bhattacharya, B.K. and Toussaint, G.T., "On geometric algorithms that use the furthest-point Voronoi diagram," *Computational Geometry*, Edited by Toussaint, G.T., North Holland, pp.43-61, 1985.

[B] Brown, K.Q., "Voronoi diagrams from convex hulls," *Information Processing letters*, Vol.9, pp.223-228, 1979.

[C] Chazelle, B.M., *Computational geometry and convexity*, Ph.D. Thesis, Yale University, 1980.

[D] Dyer, M.E., "Linear time algorithms for two- and three-variable linear programs," *SIAM Journal on Computing*, Vol.13, No.1, pp.31-45, 1984.

[DK] Dobkin, D.P. and Kirkpatrick, D.G., "A linear algorithm for determining the separation of convex polyhedra," *Journal of Algorithms*, Vol.6, pp.381-392, 1981.

[DR] Dobkin, D.P. and Reiss, S.P., "The complexity of linear programming," *Theoretical Computer Science*, Vol.11, pp.1-18, 1980.

[E] Edelsbrunner, H. *Algorithms in combinatorial geometry*, Springer-Verlag, 1987.

[F] Fisk, S., "Separating point sets by circles and the recognition of digital disks," *Pattern Analysis and Machine Intelligence*, Vol.PAMI-8, No.4, pp.554-556, 1986.

[KA] Kim, C.E. and Anderson, T.A., "Digital disks and a digital compactness measure," *Proc. of 16th ACM Symposium on Theory of Computing*, Washington DC, pp.117-124, 1984.

[M] Megiddo, N., "Linear-time algorithms for linear programming in R^3 and related problems," *SIAM Journal on Computing*, Vol.12, No.4, pp.759-776, 1983.

[OR] O'Rourke, J. and Rao Kosaraju, S., "Computing circular separability," *The Johns Hopkins University*, Technical Report No. JHU/EECS-85/05, 1985.

[PH] Preparata, F.P and Hong, S.J., "Convex hulls of finite sets of points in two and three dimensions," *Comm. of ACM*, Vol.20, pp.87-93, 1977.

[PS] Preparata, F.P. and Shamos, M.I., *Computational geometry - an Introduction*, Springer-Verlag, 1985.

[S] Shamos, M.,I., *Computational geometry*, Ph.D. Thesis, Yale University, 1978.

Computational Morphology
G.T. Toussaint (editor)

SYMMETRY FINDING ALGORITHMS

Peter Eades *

University of Queensland
St. Lucia, Queensland, Australia 4067

Several algorithms for finding symmetries of geometrical objects have recently appeared. In this paper the results are surveyed, and some problems which remain open are described.

1. Introduction

Humans recognise, remember, and enjoy symmetrical objects better than asymmetrical ones. There is a wealth of literature, spanning more than one hundred years, discussing the mechanisms by which humans detect symmetry (see [8]). Computer methods for symmetry detection have been motivated by Shape Analysis and Robotics applications (see, for example, [16]). Efficient symmetry detection algorithms have appeared in the recent surge of interest in Computational Geometry ([17], [4], [19], [21], [9], [12]). In this paper, we summarise the Computational Geometry approach, concentrating on the two dimensional case. Some open problems are described in the final Section.

Humans have an intuitive notion of symmetry, but intuition is not suitable for computer algorithms. Fortunately, the intuitive notions have been precisely expressed in the following mathematical definitions.

A *similarity* is a transformation of R^d, d=2,3, which preserves straight lines and angles; it is an *isometry*, or a *congruence*, if it preserves distances. An isometry $f : R^d \rightarrow R^d$, d=2,3, is a *symmetry* of a subset D of R^d if it preserves D setwise. If there is a nontrivial symmetry of D, then D is *symmetric*. The set of symmetries of D forms a group G_D; this group provides a precise measurement of the symmetry of D.

We restrict our attention to classes of sets D which are commonly studied in Computational Geometry, Graphics and Pattern Recognition: for example, finite sets of points, polyhedra, finite sets of line segments, and planar subdivisions (see [18]). For each of these classes, a particular set can be specified by a finite number of real numbers. For each class of input, the *Symmetry Problem* is to compute G_D from D; for example, the *Symmetry Problem for Polygons* is to compute G_D for a given polygon D. A set of generators of G_D would be sufficient output, or perhaps it would be enough to know whether G_D is trivial. However, in most cases it takes no more work to list every element of G_D. Each symmetry of D is a linear transformation, and can be represented, for example, as $d \times d$ matrix and a $d \times 1$ translation.

*Research partially supported by the Computational Geometry Laboratory at McGill University.

The possible groups of symmetries of sets in 2 and 3 dimensions have been classified: see, for example, [22]. If D is a bounded set (that is, it is possible to draw a sphere around D) then the possible symmetries are very simple. All such symmetries fix the centroid of D, and most algorithms assume that the origin has been translated to the centroid. The relevant classifications are mentioned below.

In two dimensions, if D is bounded then G_D consists of rotations about the centroid of D and reflections in lines through the centroid. If D is a finite set of points or line segments (for example, a polygon) then G_D has size at most $2n$, where n is the number of points or line segments. It is generated by either a single rotation (a cyclic group), a single reflection (a group of size 2), or a rotation and a reflection (a dihedral group).

In three dimensions there are five regular solids (the "platonic" solids): the tetrahedron, the octahedron, the cube, the icosahedron, and the dodecahedron. The tetrahedron has a group T of 12 symmetries. The octahedron and the cube have the same group W of symmetries; this group has size 24. The icosahedron and the dodecahedron have the same group P, of size 60. If D is a bounded set in three dimensions, then G_D may be a subgroup of T, W or P, or it consists of rotations about a line through the centroid of D and reflections in planes through the centroid. Again, the size of G_D is bounded by a linear function of the size of D, and G_D has a generating set containing at most one rotation, and at most one reflection.

Even if D is unbounded, G_D has a rather simple structure; see [7].

In Section 2 and 3 respectively, algorithms for two and three dimensional symmetry problems are described. Many of the algorithms have time complexity $O(n \log n)$, and for some restricted problems, linear time algorithms are given. The optimality of these algorithms is discussed in Section 4.

Closely related to symmetry are the notions of "similarity" and "congruence". Two sets C and D are *similar* if they differ by similarity, and they are *congruent* if they differ by an isometry. Thus a symmetric set is nontrivially self congruent. Given two sets C and D, the *Similarity Problem* is to find all similarities between C and D, and the *Congruence Problem* is to find all congruences between C and D.

The Congruence Problem is a subproblem of the Similarity Problem. For classes of sets where a measure of the dilation of the similarity is readily available, they are equivalent. For example, the largest side of a polygon can be found in linear time; thus to find similarities between a pair of polygons, we can "shrink" one by the ratio of their largest sides, then test for congruence.

The Symmetry Problem is a subproblem of the Congruence Problem. Conversely, all the algorithms presented here can be applied to detect congruence; in fact, several were originally proposed as congruence detection algorithms.

2. Algorithms in Two Dimensions

Algorithms for computing symmetries in two dimensions have been given for polygons ([17], [2], [6], [21]), finite sets of points ([4], [10], [12], [21]), and finite sets of line segments ([9]).

Throughout this Section we shall assume the origin has been translated to the centroid. Thus the possible symmetries are rotations about the origin and reflections in lines through the origin.

The set D of size n is represented as an n-tuple $L(D) = (l_1, l_2, ..., l_n)$. Several properties are required of this representation. The representation must be one-to-one up to congruence, that is, if $L(D)=L(D')$, then D is congruent to D'. Conversely, symmetries of D should induce permutations of $L(D)$ which are easily recognised, in a sense that is described below.

Algorithms to find symmetries in two dimensions have two basic steps:

Step 1: Construct the n-tuple $L(D)$.

Step 2: Find all permutations on $L(D)$ which are induced by symmetries of D.

The second step is implemented in a similar way in all the algorithms. Suppose that $L=l_1 l_2 \cdots l_n$ and $L'=l'_1 l'_2 \cdots l'_n$ are strings over some alphabet and $l'_i = l_{i+m}$ (index arithmetic is modulo n); then we say that L' is the *rotation* of L by m, and we write $L'=c_m(L)$.. Given L and L', we need to find all $m \in \{1,2,...,n-1\}$ such that $L'=c_m(L)$. This can be done by constructing the string LL by concatenating L with itself, then reporting all occurrences of L' in LL. The Knuth-Morris-Pratt fast string matching algorithm [15] can be used to do the string matching in linear time. Note that the linearity of this algorithm does not depend on the size of the alphabet (in this case, the alphabet is the set of real numbers).

The nature of the list $L(D)$ and the method by which it is constructed depends on the class of set D.

Polygons

Suppose that D is a polygon (not necessarily simple) whose sides are the line segments s_1, s_2, ..., s_n, in clockwise order. Manacher [17] (see also [1]) constructs $L(D)$ as follows. Let θ_i be the angle that s_i makes with the positive x-axis, and let

(1) $\alpha_i = \theta_{i+1} - \theta_i$, and for $i=1,2,...,n-1$, and $\alpha_n = \theta_1 - \theta_n + 2\pi$.

Denote the length of s_i by d_i. Then $L(D)$ is $((d_1,\alpha_1), (d_2,\alpha_2), ..., (d_n,\alpha_n))$. The construction takes linear time.

A rotation of D by $2\pi/k$ induces a rotation $c_{n/k}$ of $L(D)$. Thus using the Knuth-Morris-Pratt algorithm as described above, rotational symmetries can be found in linear time.

An adaption of the same technique can be used to find reflective symmetries. A reflection of D induces a reversal followed by a rotation of $L(D)$. Denote the reversal of $L(D)$ by $M(D)$. Thus to find such symmetries, we search for rotations c_m such that $M(D)=c_m(L(D))$. If such a rotation is found, then D has reflective symmetry in a line at an angle of ρ to the x-axis, where ρ is θ_t if $m+n$ is even, and $\theta_t + \alpha_t/2$ if $m+n$ is odd, where $t = \lceil (m+n+1)/2 \rceil$.

Thus Step 2 can be implemented in linear time, and we have a linear time algorithm for the Symmetry Problem for Polygons.

Variants of this algorithm are described by Manacher [17] and Akl and Toussaint ([1], [2], [3]) in a form suitable for testing similarity between two polygons. Note that the algorithm in [6] is incorrect because the proposed representation $L(D)$ for polygons is not one-to-one; see [21].

Points

If D is a set of points in the plane then $L(D)$ may be constructed as follows (from [10], [12]). If the origin is in D then we delete it; this does not change any symmetry properties of D. Represent each point in D in polar coordinates (r,θ), where $0 \le \theta < 2\pi$. Sort the points into *polar order*, to obtain a list $((r_1,\theta_1),(r_2,\theta_2), \ldots , (r_n,\theta_n))$, where $\theta_i \le \theta_{i+1}$ and if $\theta_i = \theta_{i+1}$ then $r_i < r_{i+1}$, for $i=1,2,\ldots,n-1$. Now let α_i be defined as in (1) above. The list $L(D)=((r_1,\alpha_1),(r_2,\alpha_2), \ldots, (r_n,\alpha_n))$ is a one-to one representation of D (up to congruence).

A rotation on D induces a rotation on $L(D)$ and so rotational symmetries can be found by the Knuth-Morris-Pratt algorithm as described above.

As for polygons, a reflection on D induces a reversal followed by a rotation. But here we need a little more care because the order of points with the same polar angle is not reversed. A *run* in $L(D)$ is a maximal sublist with the same polar angle, that is, a sublist $((r_i,\alpha_i), (r_{i+1},\alpha_{i+1}), \ldots, (r_j,\alpha_j))$ where $\alpha_i, \alpha_{i+1}, \ldots \alpha_{j-1}$ are all zero but α_{i-1} and α_j are nonzero. Note that the polar order of $L(D)$ implies that $r_i < r_{i+1} < \cdots < r_{j-1}$. The *run reversal* $M(D)$ of $L(D)$ is the list formed by reversing the order of occurrence of the runs of $L(D)$, but keeping the order within the runs unchanged. Using a stack, $M(D)$ can be formed from $L(D)$ in linear time. A reflection of D transforms $L(D)$ to a rotation of $M(D)$. Thus to find reflective symmetries, we use the Knuth-Morris-Pratt algorithm to find occurrences of $M(D)$ in the concatenation of $L(D)$ with itself.

The algorithm is dominated by the polar sorting and so the Symmetry Problem for Points in Two Dimensions can be solved in worst case time complexity $O(n \log n)$.

In practice, bucket sorting can be used to achieve linear expected time. We can use n wedge shaped buckets $B_1, B_2, \ldots, B_n$, where $B_i = \{(r,\theta): 2\pi(i-1) \le n\theta < 2\pi i\}$, as illustrated in Figure 1.

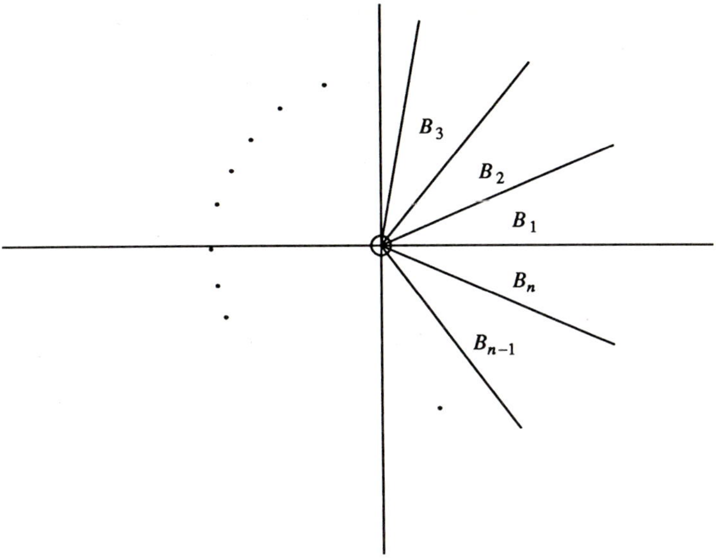

Figure 1: Wedge Shaped Buckets

In this case bucketting has some worst case advantages as well, noted in [11]. If $n=km$, $k>1$, then for $1 \le i \le k$ let S_i denote $B_{(i-1)m+1} \cup B_{(i-1)m+2} \cup \cdots \cup B_{im}$. A rotation ρ by $2\pi/k$ maps S_i to S_{i+1}, and so if ρ is a symmetry of D, then $|S_i|=m$ for $1 \le i \le k$. Thus no bucket has more than m elements, and the sorting takes (worst case) time $O(n \log(n/k))$. If ρ is a symmetry of D then $|G_D|$ is $\Omega(k)$ and thus we have an improved algorithm for finding all symmetries of D, as long as $|G_D|$ is large, that is, as long as D is highly symmetric.

Line Segments

Similar techniques are employed in [9] to find symmetries of a set of line segments.

Suppose that D is a set of line segments. Denote the midpoint of each l in D by μ_l. Note that if D is symmetrical, then the centroid of D is the centroid c of the set $\{\mu_l : l \in D\}$ of points; we translate the origin to c. Each line segment l for which $\mu_l=c$ is replaced by its two halves in D; this does not alter the symmetries of D.

For each l in D, we compute the 4-tuple (r,θ,d,β), as follows. The polar coordinates of μ_l are (r,θ) and d is the length of l. Let p be the endpoint of l which is furtherest from the origin. Then β is the angle which l makes with the line from the origin to μ_l, such that $-\pi/2<\beta\le\pi/2$, as illustrated in Figure 2.

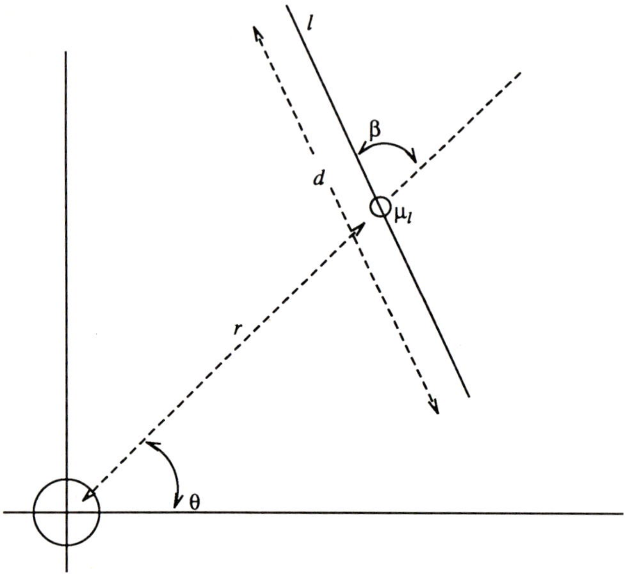

Figure 2: parameters for finding symmetries of sets of line segments.

We sort the 4-tuples on θ, r, d, and finally on the absolute value of β. If α is the difference between successive polar angles θ as for points, then the list $L(D)$ has entries (α, r, d, β). With a little care for reflections, symmetries of D can be detected as rotations, and reversal-rotations, of $L(D)$ as for polygons and for points. In fact, in [9] it is shown that symmetries of sets consisting of lines and points ("drawings") can be found in time $O(n \log n)$ using the same basic approach.

3. Three Dimensions

Three dimensional symmetry detection algorithms, presented in [4], [19], and [21], are considerably more complex than the corresponding two dimensional cases. Here a brief summary of this work is given.

The problem is considerably easier if an axis is given as part of the input. The *given axis* problem is to take a line l and a set D and compute all rotations about l and reflections in planes through l which are symmetries of D. This problem can be solved in a way that is analogous to the two dimensional algorithms.

Atkinson [4] considers a finite set D of points. Translate the origin to the centroid of D, and rotate the axes so that l coincides with the z axis. Represent each point in cylindrical coordinates (r, z, θ), then construct the list $L(D)$ as for points but where each entry of $L(D)$ has an extra coordinate z. The "polar sort" can be done on θ, r, then z. The Knuth-Morris-Pratt algorithm can be used in the same way as in Section 2 to find all rotational and reflective symmetries.

A more complex approach is used in [21] to solve the given axis problem for polyhedra with planar triply connected surface graphs in linear time. The proposed line of symmetry l can intersect the surface of the polyhedron P in only three ways: through a vertex v, through the midpoint of an edge e, or through the centroid of a face f. Let Γ_1 be the set of vertices graph-theoretically adjacent to v, at each end of e, or around f, depending on the way that l intersects P. Then let Γ_k be the set of vertices which are not in Γ_j for $j<k$ but are adjacent to vertices in Γ_{k-1}. Note that if P is symmetrical about l then each Γ_k must be arranged symmetrically about l. An algorithm is given in [21] which constructs a list $L(P)$ of the vertices of P from the Γ_k in linear time; again, the Knuth-Morris-Pratt algorithm can be used to find symmetries.

Thus in three dimensions the difficulty is in proposing possible axes of symmetry.

Atkinson [4] uses several "reductions" described below to propose axes of symmetry for finite sets of points. Suppose that Π is a geometric property which is invariant under isometries. Let D_1 be the subset of D which has property Π and let D_2 be $D-D_1$. The sets of symmetries of D_1 and D_2 are both supersets of the set of symmetries of D. If both D_1 and D_2 are nonempty then the problem of proposing axes for D is reduced to proposing axes for a set of at most half the size of D. If none (or all) of D has property Π then we can use this information is determining the structure of D.

For instance, suppose that d is the maximum distance of a point in D from the centroid. Let Π be the property of having distance d from the centroid. If there are some points of D which do not have property Π, then we search for axes of the smaller of D_1 and D_2; the problem size is reduced by at least half. If all D has property Π then D lies on a sphere of radius d about the centroid; this gives us significant information about D, which can be used to propose axes.

Four such properties Π are given in [4], and it is shown that in time $O(n \log n)$ one can obtain a set D^* such that every symmetry of D is a symmetry of D^*, and $|D^*|$ is bounded by a constant. If all the points of D^* lie on a line through the centroid, then this line is proposed as an axis for D; if not, then there are a constant number of symmetries of D^* and in time $O(n \log n)$ one can check whether each is a symmetry of D. Thus symmetries of sets of points in 3 dimensions can be found in time $O(n \log n)$.

The linear time isomorphism algorithm of Hopcroft and Wong [14] is used in [21] to propose lines of symmetry for polyhedra whose surface graphs are planar and triply connected. The Hopcroft-Wong algorithm is quite complex, and the authors note that because of the large constant involved, it may not be practical. However, it does succeed in showing that symmetries of polyhedra with triply connected planar surface graphs can be found in linear time.

Sugihara [19] has a different approach for polyhedra. The planar graph isomorphism algorithm of Hopcroft and Tarjan [13] is slightly slower ($O(n \log n)$) than the Hopcroft-Wong algorithm. However, it can be applied to a wider class of graphs; further, it is considerably simpler. Sugihara shows how to modify the Hopcroft-Tarjan algorithm to find all symmetries of a wide class of polyhedra in time $O(n \log n)$.

4. Optimality

Several authors ([4], [19], [21], [12], [9]) have pointed out that many Symmetry Problems require time $\Omega(n \log n)$ on fixed degree decision tree machines (see [5]). In fact the *Reflection Test Problem* below has an $\Omega(n \log n)$ lower bound.

Reflection Test Problem:

INSTANCE: A set X of points on the x-axis.

QUESTION: Is reflection in the y-axis a symmetry of X?

There is a linear time transformation (see [18]) from the *Set Equality Problem* (testing whether two sets of positive integers are equal) to the Reflection Test Problem as follows. Suppose that $S=\{s_1, s_2, ..., s_n\}$ and $T=\{t_1, t_2, ..., t_n\}$ are sets of positive integers; let X denote the set $\{s_1, s_2, ..., s_n, -t_1, -t_2, ..., -t_n\}$, considered as a set of points on the x-axis. Reflection in the y-axis is a symmetry of X if and only if $S=T$. The Set Equality Problem has an $\Omega(n \log n)$ lower bound for fixed degree decision tree machines (see [5] or [18]). Note that a half-turn (rotation by π) has the same effect on X as reflection in the y-axis, so the lower bound applies to rotations as well. Symmetry Problems which have the Reflection Test Problem as a subproblem have an $\Omega(n \log n)$ lower bound for fixed degree decision tree machines; this includes Symmetry Problems for points and for line segments in two and three dimensions. A more refined argument gives a similar result for a general class of polyhedra [21].

With one exception, the algorithms in Sections 2 and 3 can be performed on a fixed degree decision tree machine by using the usual tricks of comparing angles and distances without explicitly computing them. Thus for this machine model, they are optimal. The one exception is the bucket sorting technique described in Section 2. This needs a more powerful machine: a "real RAM" (see [18]) with the ability to perform truncation of real numbers in constant time.

5. Final Remarks: "Near" Symmetry

Although the definition of symmetry in Section 1 is appropriate for most circumstances, it does not quite capture all the human intuition of symmetry: for instance, the following diagrams appear to have some "symmetry" (in the intuitive sense), but according to our precise definition, have none.

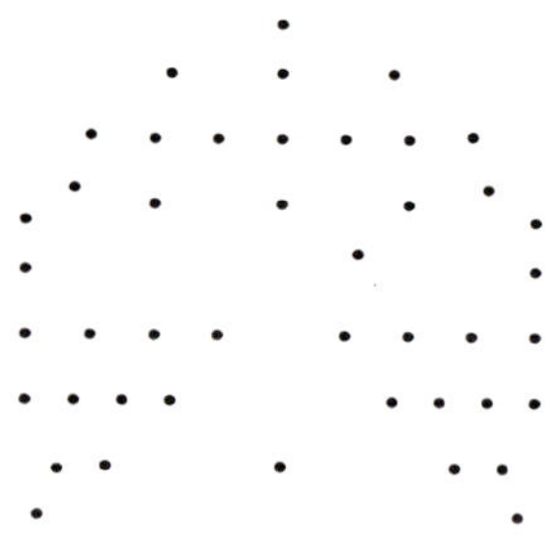

Figure 3: nearly symmetric set of points.

Figure 4: a drawing that can be decomposed into two symmetric drawings.

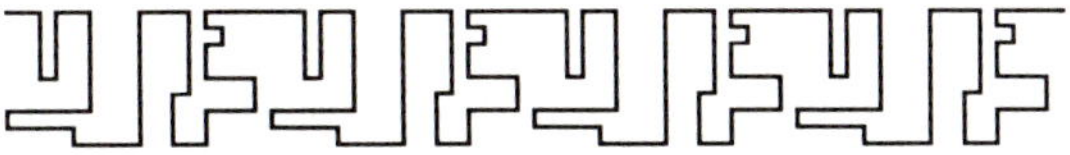

Figure 5: a finite strip pattern.

The set of points in Figure 3 has a large symmetric subset. The drawing of Figure 4 can be decomposed into two subsets, each of which is symmetric. The "near" symmetry of Figure 5 is not so easily categorized: it seems that the translation in the horizontal direction is "nearly" a symmetry (it would be a translational symmetry if the pattern was repeated infinitely).

The first example suggests the following problem:

Maximal Symmetric Subset Problem (MSS) : Given a set D in R^d (d=2,3), find a symmetric subset of maximal cardinality.

Note that MSS can also be stated as the problem of finding a symmetric superset of D of minimal cardinality.

Because a symmetry is fixed by its action on a small number of points, MSS can be solved in polynomial time. For instance, for each triple (p,q,r) of distinct points there is at most one rotation of the plane which maps p to q and q to r. Thus there are at most $n(n-1)(n-2)$ candidates for a rotational symmetry of three or more points in D. It is not difficult to compute the size of the subset of D which is fixed by a given rotation in time $O(n \log n)$. A similar result can be applied to reflections. Hence a maximal symmetric subset can be found in time $O(n^4 \log n)$. It would be interesting to know if there is a faster algorithm than this naive approach.

Figure 4 suggests two problems:

Minimal Symmetric Decomposition Problem (MSD) : Given a set D in R^d (d=2,3), find a set $\{D_1, D_2, ..., D_k\}$ of symmetric subsets of D such that the union of the D_i is D and k is minimal.

Minimal Symmetric Partition Problem (MSP) : Given a set D in R^d (d=2,3), find a set $\{D_1, D_2, ..., D_k\}$ of disjoint symmetric subsets of D such that the union of the D_i is D and k is minimal.

The complexity of MSD and MSP is open. It would be particularly interesting to discover whether either is NP-hard.

Figure 5 is reminiscent of the considerable mathematical literature about "strip patterns", that is, infinite sets in the plane, bounded above and below in the y direction, and invariant under a translation in the x direction. (See, for example, [20].) This literature deals with drawings such as Figure 5 by imagining the pattern as being extended to infinity in both directions. Such imagination, however, is beyond the scope of Computational Geometry. A precise mathematical model which can deal with finite strip patterns and their "symmetries" could lead to interesting algorithms for their recognition.

References

[1] Akl, S.G., Comments on: G. Manacher, An application of pettern matching to a problem in geometrical complexity, *Information Processing Letters*, 7, (1978) 86.

[2] Akl, S.G. and Toussaint, G.T., An Improved Algorithm to Check for Polygon Similarity, *Information Processing Letters*, 7, (1978) 127-128.

[3] Akl, S.G. and Toussaint, G.T., Addendum to an Improved Algorithm to Check for Polygon Similarity *Information Processing Letters*, 8, (1979) 157-158

[4] Atkinson, M.D., An Optimal Algorithm for Geometrical Congruence, *Journal of Algorithms* 8 Number 2 (1987), 159-172.

[5] Ben-Or M., Lower Bounds for Algebraic Computation Trees, Proceedings of the Fifteenth Symposium on the Theory of Computing, Boston, 80-86, 1983.

[6] Bykat, A., On polygon similarity, *Information Processing Letters*, 9, (1979) 23-25.

[7] Coxeter, H.M., *Introduction to Geometry*, Wiley, 1961.

[8] Eades, P. and Lee A., Perception of Symmetry, Technical Report Number 52, Department of Computer Science, University of Queensland, 1985.

[9] Eades, P. and Ng H.C., An algorithm for detecting symmetries in drawings, *Ars Combinatoria* 23A (1987), 95-104.

[10] Eades, P., Urrutia, J. and Santoro, N., Finding Symmetries of Sets of Points in the Plane, unpublished manuscript, 1984.

[11] Eades, P., A Worst Case Improvement for Symmetry Detection by Bucketting, Technical Report Number 83, Department of Computer Science, University of Queensland, 1987.

[12] Highnam, P.T., Optimal algorithms for finding the symmetries of a planar point set, *Information Processing Letters* 22 (1986) 219-222.

[13] Hopcroft, J.E. and Tarjan, R.E., A V log V algorithm for isomorphism of triconnected planar graphs, *Journal of Computer and System Sciences*, 7 (3) (1973), 323-331.

[14] Hopcroft, J.E. and Wong, J.K., Linear Time Algorithm for Isomorphism of Planar Graphs, Proceedings of the Sixth Annual ACM Symposium on Theory of Computing, Seattle, (1974), 172-184.

[15] Knuth, D., Morris, J. and Pratt, V., Fast Pattern Matching in Strings, *SIAM Journal of Computing*, 6, (1977) 322-350.

[16] Leou, J-J. and Trai W-H., Automatic Rotational Symmetry Determination for Shape Analysis, *Pattern Recognition* 20 No. 6 (1987), 571-182.

[17] Manacher, G., An Application of Pattern Matching to a Problem in Geometric Complexity, *Information Processing Letters*, 2 (1976), 6-7.

[18] Preparata, F.P. and Shamos, M.I., *Computational Geometry: An Introduction*, Springer Verlag, 1985.

[19] Sugihara, K., An $n \log n$ algorithm for Determining the Congruity of Polyhedra, *Journal of Computer and System Sciences* 29 (1984), 36-47.

[20] Weyl, H., *Symmetry*, Princeton University Press, 1951.

[21] Wolter, J.D., Woo, J.D., and Volz R.A., Optimal algorithms for symmetry detection in two and three dimensions, *The Visual Computer*, 1 (1985), 37-48.

[22] Paul B. Yale, *Geometry and Symmetry*, Holden-Day, 1968.

Computational Morphology
G.T. Toussaint (editor)

COMPUTING THE RELATIVE NEIGHBOUR DECOMPOSITION OF A SIMPLE POLYGON

Hossam A. ElGindy and *Godfried T. Toussaint*

School of Computer Science, McGill University,
Montreal, Quebec H3A 2K6, Canada.

In *computational geometry* one may be interested in decomposing a polygon into simpler components, monotone polygons for example, in order to solve the geometric problem at hand more efficiently. However, in *pattern recognition*, where the motivation is *morphological*, one is interested in decomposing a polygon into perceptually meaningful parts. Therefore we can relax the strict requirement that the components be of a certain form such as convex or monotone and we can investigate decompositions which are *procedure oriented* rather than *component oriented.* In this paper we study the properties of a procedure oriented decomposition, termed the *relative neighbour decomposition*, and present different algorithms for performing such a decomposition.

1. INTRODUCTION

The decomposition of a simple polygon into simpler components plays an important role in syntactic pattern recognition. Some examples of possible decompositions are: decompositions into convex polygons [1-3], decompositions into spiral polygons [4], decompositions into monotone polygons [5], and decompositions into star-shaped polygons [6]. A survey of these methods is contained in [7,8]. The objective of such decompositions is to obtain perceptually meaningful components. Therefore we can relax the strict requirement that the components be of a certain form, and we can investigate decompositions which are *procedure oriented* rather than *component oriented.*

In this paper we propose a *procedure directed* decomposition, termed the *relative neighbour decomposition* (RND), that uses the idea of the *relative neighbourhood graph* (RNG) of a set of points to determine which diagonals of a simple polygon P should be added to form the RND(P). We then exhibit a proof that the RND is planar (i.e., the simpler components of P do not overlap). Section 4 is devoted to describing different algorithms for computing the RND of a simple polygon. We finally conclude with some open problems in section 5.

2. THE RND PROBLEM

Consider a set of n distinct points in the plane $P = p_1, p_2, \ldots, p_n$ given by cartesian coordinates. The lune of two points p_i and p_j, denoted by LUNE(p_i, p_j) is defined as the intersection of two circles with radii equal to $d(p_i, p_j)$ centered at p_i and p_j, where d denotes euclidean distance. The RNG is a planar graph on P with the property that $\overline{p_i\, p_j}$ is an edge of the RNG if and only if no other point lies in the interior of LUNE(p_i, p_j) [9]. The straightforward approach to computing the RNG leads to an $O(n^3)$ algorithm, but in [9] an $O(n^2)$ algorithm is presented which uses the Voronoi diagram. An $O(n \log n)$ algorithm was exhibited in [10] but it is easy to show that it does not always work correctly [11]. Supowit [12], in a detailed study of the RNG, presented an $O(n \log n)$ algorithm for computing the RNG of a planar set and $O(n)$ algorithm for the RNG of a convex polygon. O'Rourke [13] and Lee [14] considered the RNG under non-euclidean metrics, and the *discrete* RNG is applied to image analysis in [15]. Algorithms with fast *expected* running time are given in [16,17]. The RND in higher dimensions is considered in [18].

The *relative neighbour decomposition* (RND) of a simple polygon P is obtained by extending the definition of the RNG as follows: two vertices of P, say p_i and p_j, are joined by a diagonal if and only if 1) p_i and p_j are *visible*, where two vertices of P are *visible* if and only if the diagonal joining them lies in P, and 2) no other *visible* vertex from p_i or p_j lies inside LUNE(p_i, p_j). Figure 2.1 shows a simple polygon and its RND.

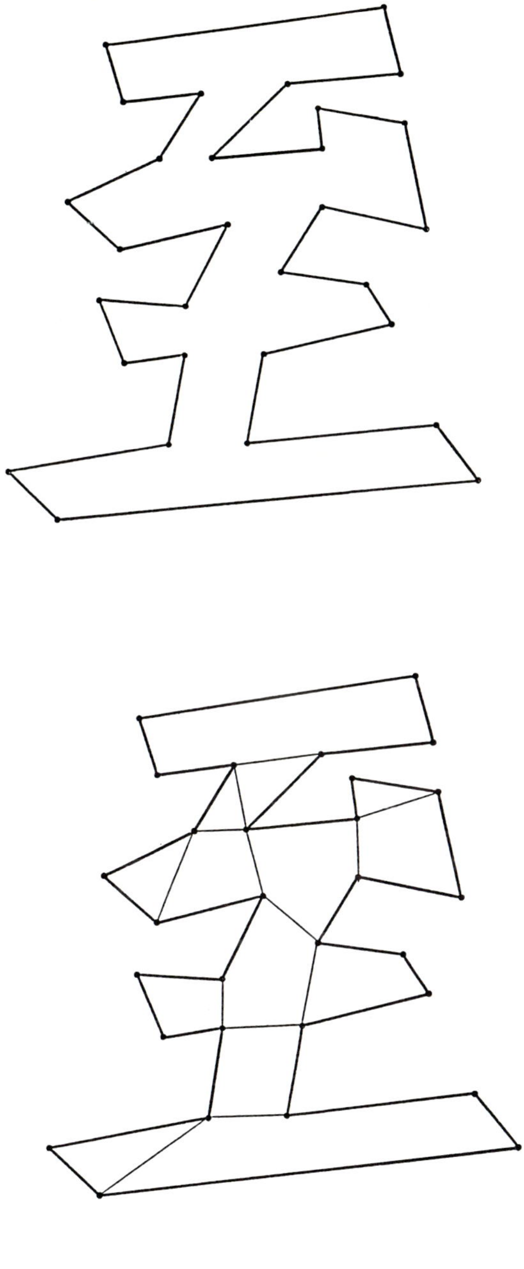

Fig. 2.1

A graph which is closely related to the RNG is the *Gabriel* graph. The disk of p_i and p_j, denoted by DISK(p_i, p_j), is defined as follows:

$$\text{DISK}(p_i, p_j) \equiv \{\ x:\ [d^2(x,p_i)+d^2(x,p_j)]^{1/2} \leq d(p_i,p_j)\ \}$$

Clearly, p_i and p_j lie on a circle and determine its diameter. This circle is known as the "circle of influence" and is an alternate definition of relative neighbourhood to that specified by the "lune of influence" used in the RNG. Since the disk is a subset of the lune it follows that the RNG is a subgraph of the Gabriel graph (GG). Matula and Sokal [19] derive many interesting properties of Gabriel graphs and show how it can be computed in $O(n\log n)$ time using the Voronoi diagram. The *Gabriel graph decomposition* (GGD) of a simple polygon P can be obtained by extending the definition of the GG as follows: two vertices of P, say p_i and p_j, are joined by a diagonal if and only if 1) p_i and p_j are *visible* and 2) no other *visible* vertex from p_i or p_j lies inside DISK(p_i, p_j). Such a decomposition yields many more components than the RND. While it is interesting in its own right, it does not appear to be as useful as the RND for pattern recognition applications, where we want the decomposition to contain a relatively small number of components.

It is known that the RNG and the GG of a set of points are planar graphs [9,19]. But it does not follow that the RND and GGD of polygon are planar, since the latter graphs may contain edges with non-empty lunes and disks. In other words, the RND (GGD) of a polygon may contain edges not contained in the RNG (GG) of its vertices. Thus the first question that arises concerning these decompositions is their planarity. In the following section we show that the GGD is planar. It follows trivially that if the GGD is planar then so is the RND.

3. PROOF OF PLANARITY

First we introduce some notation. An edge or line segment with end points A and B will be denoted by $\overline{AB}$. A triangle with vertices A, B and C will be denoted by $\Delta(ABC)$ whereas *angle*(ABC) will denote the angle at vertex B formed by vertices A and C. Let bd(DISK(A,B)) and int(DISK(A,B)) denote the boundary and interior of DISK(A,B) respectively.

Lemma 3.1 If the bd($\Delta(ABC)$) lies in a simply connected region R, then int(ΔABC)) also lies in R.

Proof. Let $\overline{AB}$, $\overline{BC}$, and $\overline{AC}$ all lie in R and let y be a point interior to $\Delta(ABC)$ and exterior to R. It follows from Jordan's curve theorem that bd(R) must intersect bd($\Delta(ABC)$) at least twice, say at a and b. It follows that portions of bd($\Delta(ABC)$) lying between a and b must lie in the exterior of R which is a contradiction, thus proving the lemma. ■

Lemma 3.2 Any triangle is contained in the union of the disks of any two of its sides.

Proof. Without loss of generality, consider $\Delta(ABC)$ with DISK(A,B) and DISK(B,C) as the two chosen sides, and refer to Figure 3.1. The bd(DISK(A,B)) and bd(DISK(B,C)) intersect at B and D. Draw a line from B to D. Since $\overline{BC}$ is the diameter of DISK(B,C) and D lies on bd(DISK(B,C)), it follows that $angle(BDC)=\frac{\pi}{2}$. Similarly, $angle(ADB)=\frac{\pi}{2}$. Therefore, $\overline{AD}\cup\overline{DC}=\overline{AC}$. Since $\overline{DC} \subset$ DISK(B,C) and $\overline{AD} \subset$ DISK(A,B), it follows that $\overline{AC}\subset$\{DISK(A,B)$\cup$DISK(B,C)\}. Furthermore, by definition, $\overline{AB} \subset$ DISK(A,B) and $\overline{BC} \subset$ DISK(B,C). Since all three sides of $\Delta(ABC)$ are contained in the union of the two disks, it follows from lemma 3.1 that $\Delta(ABC) \subset$ \{DISK(A,B) $\cup$ DISK(B,C)\}. ■

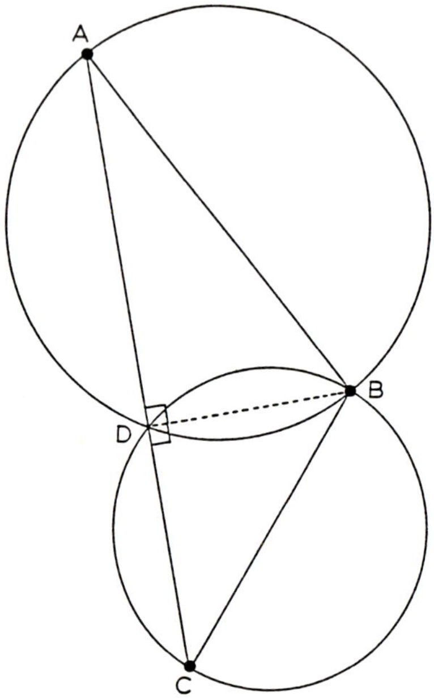

Fig. 3.1

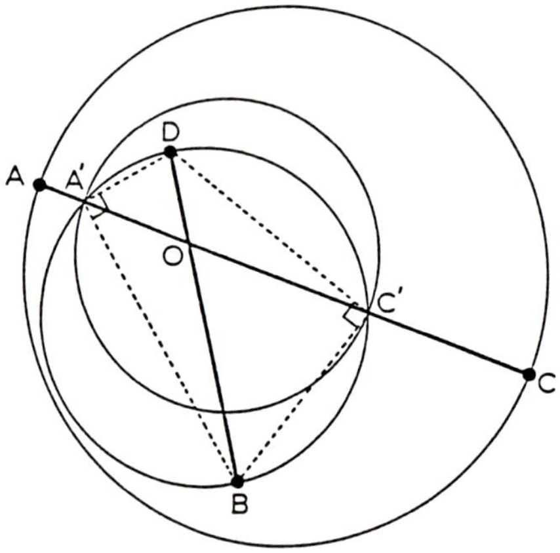

Fig. 3.2

Theorem 3.1 The relative neighbour and the Gabriel decompositions of a simple polygon P are planar (i.e., no two diagonals of the RND or the GGD intersect each other or the edges of P).

Proof. We begin by assuming that two given diagonals of the GGD, $\overline{AC}$ and $\overline{BD}$, intersect and then proceed to show that this leads to a contradiction. Two cases arise: either all vertices A, B, C, and D are visible from each other (Case A), or at least two of the vertices are not visible to each other (Case B). Note that $\overline{AC}$ and $\overline{BD}$ do not intersect polygon edges.

Case A: Consider diagonal $\overline{DB}$ and DISK(D,B). The position of the remaining two vertices A and C with respect to DISK(D,B) determines four possible situations: both A and C lie in DISK(D,B), only A does, only C does, and finally neither of the two points lie in DISK(D,B). In the first three cases, at least one visible vertex of either B or D lies in DISK(D,B) and hence $\overline{DB}$ cannot be a diagonal of the GGD, a contradiction. For the case where both A and C lie outside of DISK(D,B) refer to Figure 3.2. Since $\overline{AC}$ intersects $\overline{DB}$ and $DB \subset$ DISK(D,B), it follows that $\overline{AC}$ intersects DISK(D,B). Let A' and C' denote the points where $\overline{AC}$ intersects DISK(D,B) and construct DISK(A',C'). Points A',D,C',B determine a quadrilateral inscribed on bd(DISK(D,B)). Since $\overline{DB}$ is the diameter of DISK(D,B), and A' and C' lie on its boundary, we have $angle(DA'B) = \frac{\pi}{2}$ and $angle(DC'B) = \frac{\pi}{2}$. Since the sum of the angles of a convex n-gon is (n-2)π, it follows that $angle(A'DC') + angle(A'BC') = \pi$, and either one of these angles must be greater than $\frac{\pi}{2}$ or both must equal $\frac{\pi}{2}$. In either case, since $\overline{A'C'}$ is the diameter of DISK(A',C'), at least one of the two vertices D,B (the one whose angle is greater than $\frac{\pi}{2}$) must lie in DISK(A', C'). Since $\overline{A'C'} \subset \overline{AC}$, we have DISK($A'$,$C'$) $\subset$ DISK(A,C). Therefore at least one of the two vertices D,B must lie in DISK(A,C). Since both D and B are visible from A and C, $\overline{AC}$ cannot be a diagonal of the GGD, a contradiction.

Case B: If not all the vertices are visible from each other, then at least one pair of vertices of the set {(A,B) , (B,C) , (C,D) , (D,A)} is an invisible pair. Without loss of generality, let A and B be one of the invisible pairs, and refer to Figure 3.3. The intersection of $\overline{AC}$ and $\overline{BD}$ is denoted by O. Since $\overline{DO} \subset \overline{DB}$ it follows that DISK(D,O) $\subset$ DISK(D,B). Similarly, DISK(A,O) $\subset$ DISK(A,C). >From lemma 3.2 we have that $\Delta(AOD) \subset [DISK(A,O) \cup DISK(O,D)]$. Therefore, $\Delta(AOD) \subset [DISK(A,C) \cup DISK(B,D)]$. We will now show if A and D are not visible to each other there always exists at least one vertex of P that is visible from either A or D and that lies in DISK(A,C) $\cup$ DISK(B,D).

Since (A,C) and (B,D) are visible pairs, neither of the two polygonal chains (A,...,D) of P intersects $\overline{AO}$ or $\overline{OD}$. Now, if A and D are not visible, then at least one of the chains (A,...,D) must be either completely contained in $\Delta(AOD)$ or intersect $\overline{AD}$ at least once. With $\overline{DA}$ as the x-axis and D the origin, let UCH be the *upper convex hull* of vertices of the polygon that lie in int($\Delta(AOD)$), Figure 3.4. On this convex chain, let p_i be the vertex adjacent to A and p_j be the vertex adjacent to D. Since $(\overline{DO} \cup \overline{OA}) \subset$ P, the polygon $(D,O,A,p_i,...,p_j) \subset$ P, and no other part of the boundary of P lies in $(D,O,A,p_i,...,p_j)$. Therefore, $\overline{Dp_j} \subset$ P and $\overline{Ap_i} \subset$ P and therefore p_i is visible from A whereas p_j is visible from D.

We have thus far established that there must exist vertices in $\Delta(DOA)$, and hence in [DISK(A,C) $\cup$ DISK(B,D)], that are visible from A or D. It remains to show that at least one of the two vertices p_i, p_j is contained in the disk determined by the vertex from which it is visible. Two possibilities arise for p_i and p_j. Either they are one and the same vertex or they are distinct. In the first case, it must lie in one of the the disks, DISK(A,C) or DISK(B,D), and is visible from both A and D. Therefore one of the edges $\overline{AC}$ or $\overline{BD}$ cannot be a diagonal of the GGD of P, a contradiction. In the second case, we have two subcases: (a) both p_i and p_j lie in at least one of the disks, or (b) p_i lies in one disk and p_j in the

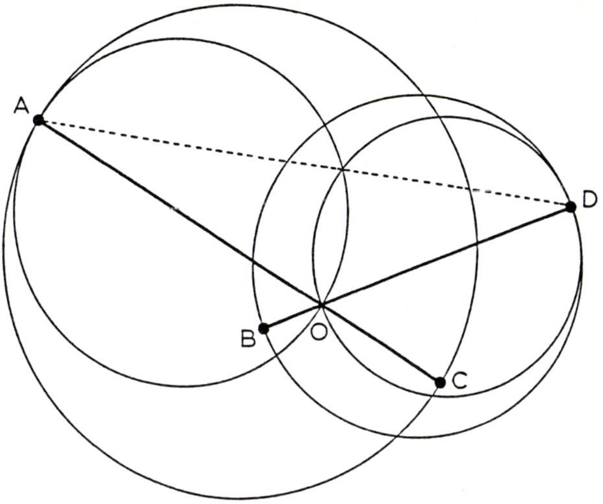

Fig. 3.3

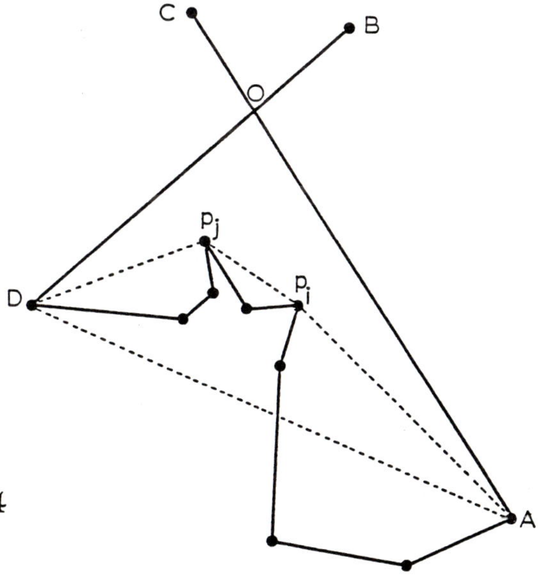

Fig. 3.4

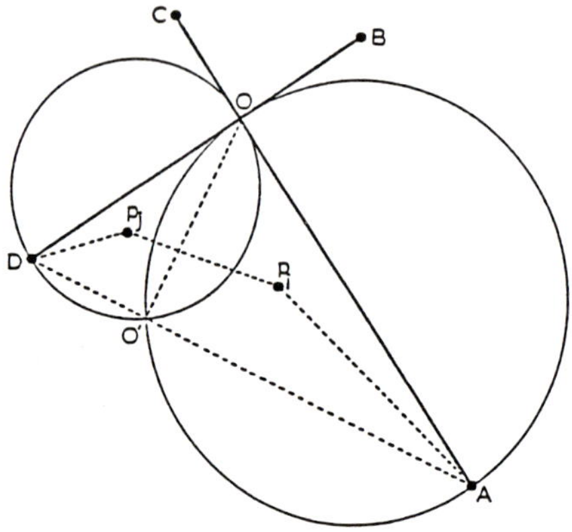

Fig. 3.5

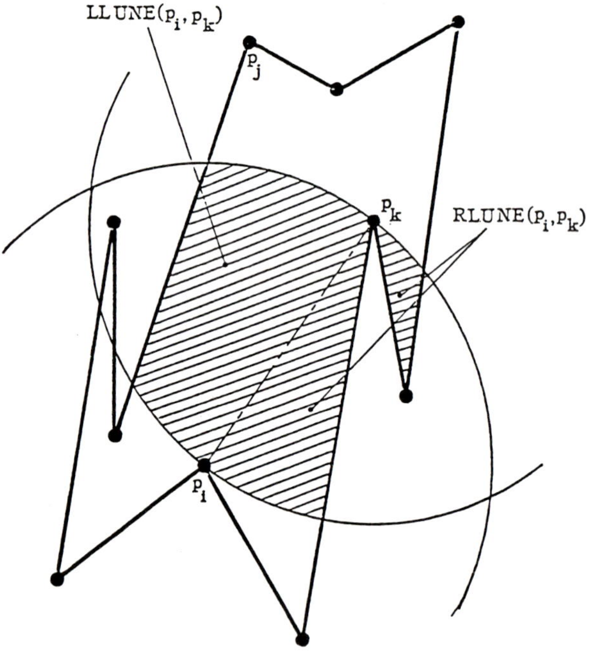

Fig. 4.1

other. In subcase (a) where both p_i and p_j lie in at least one of the disks. Without loss of generality, assume that they lie in DISK(A,C). One of the two vertices is visible from A. Therefore $\overline{AC}$ cannot be a diagonal of the GGD of P, a contradiction. In subcase (b) we must show that the situation where $p_i \subset$ DISK(B,D) and not DISK(A,C), and correspondingly for p_j, cannot arise. Let the intersection of bd(DISK(O,D)) and bd(DISK(O,A)) with $\overline{AD}$ be denoted by O' and refer to Figure 3.5. Since the chain UCH is convex, it is monotonic with respect to the line passing through $\overline{AD}$. If p_i were outside DISK(A,C) and p_j were outside DISK(B,D), then $p_i \subset \text{int}(\Delta(DOO'))$ and $p_j \subset \text{int}(\Delta(AOO'))$. Therefore the chain UCH must intersect $\overline{OO'}$ more than once. Since $angle(OO'A) = \frac{\pi}{2}$, then the chain UCH is not monotone with respect to the line passing through $\overline{AD}$, a contradiction. Therefore, p_i lies in DISK(A,C) and p_j lies in DISK(B,D) and neither $\overline{AC}$ nor $\overline{BD}$ can be diagonals of the GGD of P, a contradiction.

This completes the proof that no two diagonals can intersect each other or any edge of P and therefore the GGD(P) is planar. ■

4. ALGORITHMS

In this section we describe three algorithms for computing the relative neighbour decomposition of a simple polygon P. The algorithms for the Gabriel neighbour decomposition are similar.

A straight forward approach is to consider each pair of vertices, determine whether they are visible, find all other vertices visible from either of them, and then perform the LUNE inclusion tests. This approach is presented as Algorithm RND-1 below.

Algorithm RND-1

For all pairs of vertices p_i, p_j, $i, j = 1,2,...,n$, $i<j$, do the following:

Step 1 Determine whether $\overline{p_i p_j} \subset P$. If yes, GOTO Step 2 with a candidate diagonal; ELSE select a new pair of vertices and GOTO Step 1.

Step 2 For each vertex p_k of P, $k = 1,2,...,n$, $k \neq i \neq j$: determine whether at least one of the edges $p_k\, p_i$ or $p_k\, p_j$ lies in P. Those that do form the set V_{ij} of vertices visible from either p_i or p_j.

Step 3 For each vertex p_m of V_{ij} determine whether p_m lies in the interior of LUNE(p_i, p_j); if no vertex of V_{ij} lies in the interior of the lune then $\overline{p_i\, p_j}$ forms a diagonal of the RND(P); ELSE it does not.

Determining whether an edge lies in P can be done in $O(n)$ time. Thus Step 1 can be done in $O(n)$ time. In Step 2, $O(n)$ calculations are performed for each of the n-2 vertices. Hence Step 2 runs in $O(n^2)$ time for each candidate diagonal determined in Step 1. Finally, Step 3 can be carried out in $O(n)$ time. In the worst case, $O(n^2)$ candidate diagonals are processed in Step 2. Thus algorithm RND-1 is dominated by Step 2 and has a worst-case running time of $O(n^4)$.

The *visibility graph* VG(P) of a polygon P as defined by Shamos [20] is a graph whose nodes represent the vertices of P and contains an edge between two nodes if and only if the corresponding vertices in P are visible. Shamos [20] proposed an algorithm for computing VG(P) in $O(n^2)$ time, but the algorithm had been found to fail in some cases [21]. However, VG(P) can be computed in $O(n^2)$ time using one of the algorithms in [21] or [22] for computing the visibility polygon from a point. Algorithm RND-2 below uses the algorithm in [21] or [22] for computing the VG(P) to obtain the relative neighbour decomposition of P.

Algorithm RND-2

Step 1 Compute VG(P).

Step 2 For each edge of VG(P), say $\overline{p_i\, p_j}$, determine whether any neighbours of p_i and p_j in VG(P) lie in the in the interior of LUNE(p_i,p_j); if not, then $p_i\, p_j$ forms a diagonal of the RND(P).

Let e denote the number of edges in VG(P), i.e., e equals n plus the number of pairs of vertices in P that are candidate diagonals for the RND(P). Step 1 can be performed in $O(n^2)$ time using one of the algorithms in [21] or [22]. In Step 2, for each edge of the e edges in VG(P) at most n-2 vertices of P must be checked for lune inclusion. Therefore, Step 2 runs in $O(en)$ time. Clearly $e \leq n(n-1)/2$, where equality is attained for convex polygons. Hence, $O(n) \leq e \leq O(n^2)$. Thus algorithm RND-2 is dominated by Step 2 and has a complexity of $O(ne)$. In the worst case $e = O(n^2)$, and thus the algorithm runs in $O(n^3)$ time. However, in practice this algorithm would most likely be applied to patterns where e is closer to $O(n)$ and we can expect the algorithm then to run in less than $O(n^3)$ time.

A different approach is to consider each vertex of the polygon separately, and report *all* diagonals incident on that vertex which are candidates for the RND(P). Diagonals selected as candidates by *both* end-points form the relative neighbourhood decomposition of P. This approach is presented as algorithm RND-3 below.

Algorithm RND-3

Step 1 For each vertex p_i, $i = 1,2,...,n$, do the following:

1.A. report the set V_i of vertices visible from p_i sorted in clockwise angular order around p_i.

1.B. report the element $s \in V_i$ with the smallest euclidean distance to p_i. Note that no other element of V_i lies in the interior of LUNE(s, p_i).

1.C. scan the elements of V_i in clockwise order, starting at s, as follows:

```
p ← s ; q ← element in V_i following s
while q ≠ s do
   if p = q then
      q ← element in V_i following q
   else
      if p ∈ LUNE(q, p_i) then
         mark the vertex q
         q ← element in V_i following q
      else
         p ← element in V_i following p
end while
```

1.D. scan the elements of V_i in counterclockwise order, starting at s, in a similar fashion to step 1.C.

1.E. For each vertex in V_i: determine whether it has been marked in either of the previous scans. Those vertices that have not been marked form with the vertex p_i the set RN_i of candidate diagonals incident on p_i.

Step 2 For each edge $\overline{p_i\, p_j}$, $i,j = 1,2,...,n$, do the following:

if $\overline{p_i\, p_j}$ is a member of both RN_i and RN_j *then*

report the edge as an RND(P) diagonal

Reporting the set of visible vertices from each vertex of the polygon and then selecting the closest one can be performed in $O(n)$ time. Scanning the visible vertices to mark the diagonals incident on a certain vertex that cannot be RND(P) diagonal is also performed in $O(n)$ time. Thus Step 1 runs in $O(n)$ time for each of the n vertices of the polygon. It is not hard to see that Step 2 can be implemented to run in $O(n^2)$ time. Therefore, algorithm RND-3 has a worst-case running time of $O(n^2)$. Correctness of Step 1 follows from the correctness of the used hidden-line algorithm [21,22] and from the following lemmas.

Lemma 4.1 Let p_j and p_k be two vertices in V_i, and let L_{ij} be the subset of V_i that lie in the directed angle $(p_k\ , p_i\ , p_j)$, as shown in Figure 4.1. Note that the angle lies to the left of the p_k, p_i, p_j.
If LUNE(p_i, p_j) is empty of vertices in L_{ij}, then the part of LUNE(p_i, p_k) to the left of the directed line joining p_i to p_k, denoted by LeftLUNE(p_i, p_k), is empty of vertices in L_{ij} or contains p_j.

Proof. There are four possible cases which we consider separately:

case 1 $d(p_i\ , p_k) < d(p_i\ , p_j)$:

the part of LeftLUNE(p_i, p_k) to the left of the directed line joining p_i to p_j, if any, lies in the interior of LUNE(p_i, p_j).

case 2 $d(p_i\ , p_k) \geq d(p_i\ , p_j)$ and the directed angle $(p_j\ , p_i\ , p_k) < \frac{\pi}{3}$:

p_j lies in the interior of LeftLUNE(p_i, p_k).

case 3 $d(p_i\ , p_k) \geq d(p_i\ , p_j)$ and $\frac{\pi}{3} \leq$ the directed angle $(p_j\ , p_i\ , p_k) \leq \frac{\pi}{2}$:

If p_j lies in the interior LeftLUNE(p_i, p_k), the lemma is satisfied. If p_j lies in the exterior of LeftLUNE(p_i, p_k), the part of LeftLUNE(p_i, p_k) to the left of the directed line joining p_i to p_j, if any, lies in the interior of LUNE(p_i, p_j) due to the monotonicity of the boundary of the lune.

case 4 the directed angle $(p_j\ , p_i\ , p_k) > \frac{\pi}{2}$:

intersection of LeftLUNE(p_i, p_k) with half the plane to the left of the directed line joining p_i to p_j is empty, and is therefore empty of vertices in L_{ij}.

That completes the proof. ■

Lemma 4.2 After the algorithm RND-3 processes the vertices visible from p_i, lunes corresponding to the edges in RN_i are empty of vertices in V_i.

Proof. Prior to processing p_w in Step 1.C: let p_u be the last unmarked vertex in V_i, and F

be the subsequence $\{p, \cdots, p_{w-1}\}$ of V_i where the variable p is defined by the algorithm. There are two different cases which we consider separately:

case 1 $p = p_u$:

By lemma 4.1, we do not need to consider the vertices preceding p_u for inclusion in LeftLUNE(p_i, p_w). Therefore, the algorithm scans only the vertices of F in Step 1.C.

case 2 $p = p_v$, where $u < v < w$:

In the previous iteration, p was the vertex found to lie in LUNE(p_i, p_{w-1}) and thus p_{w-1} has been marked. In other words, each vertex in the subsequence $\{s, \cdots, p_{v-1}\}$ lies in the exterior of at least one LeftLUNE corresponding to a vertex in the subsequence $\{p_v, \cdots, p_{w-1}\}$. Therefore, by lemma 4.1, we do not need to consider the vertices preceding p_v for inclusion in LeftLUNE(p_i, p_w).

That completes the proof of Step 1.C. Correctness of Step 1.D can be shown in a similar fashion. ■

After the execution of Step 2 in algorithm RND-3, lunes of the reported diagonals are empty of vertices visible from either of their end points. Lunes corresponding to the edges marked in Steps 1.C or 1.D contain at least one vertex that is visible from either of their end points. Therefore, we conclude that algorithm RND-3 computes the relative neighbourhood decomposition of the input polygon correctly. We can now state the main result of this section.

Theorem 4.1 The relative neighbourhood decomposition of a simple polygon RND(P) can be computed in $O(n^2)$ time.

5. CONCLUDING REMARKS

In section 2 it was pointed out that the RNG of a set of points can be computed in $O(n \log n)$ time [12]. Furthermore, the RNG of a convex polygon can be computed in linear time [12]. It would thus be surprising if the $O(n^2)$ time algorithm for the RND problem of a simple polygon could not be improved. It remains to be shown whether $O(n^2)$ time is a lower bound for the problem or whether a better algorithm exists. A second open problem concerns the components of the RND. >From Figure 2.1(b) it is clear that the decomposition tends to yield convex components although not always, as evidenced by the non-convex pentagon at the lower right. It would be nice to find some characterization of these components and relate them to other well known classes of polygons [23]. Finally, on the empirical side, experiments need to be carried out to determine how well this decomposition compares with others for pattern recognition problems.

REFERENCES

[1] T. Pavlidis, "Representation of figures by labelled graphs," *Pattern Recognition* **4** (1972), pp. 5-17.

[2] B. Chazelle and D. Dobkin, "Decomposing a polygon into its convex parts," *Proceedings of the 11th ACM Symposium on Theory of Computing* (1979), pp. 38-48.

[3] B. Schacter, "Decomposition of polygons into convex sets," *IEEE Transactions on Computers* **C-27** (1978), pp. 1078-1082.

[4] H. Feng and T. Pavlidis, "Decomposition of polygons into simpler components: Feature generation for syntactic pattern recognition," *IEEE Transactions on Computers* **C-24** (1975), pp. 636-650.

[5] D.T. Lee and F.P. Preparata, "Location of a point in a planar subdivision and application," *SIAM J. Computing* **6** (1977), pp. 594-606.

[6] D. Avis and G.T. Toussaint, "An efficient algorithm for decomposing a polygon into star shaped polygons," *Pattern Recognition* **13** (1981), pp. 395-398.

[7] G.T. Toussaint, ''Pattern recognition and geometrical complexity,'' *Proc. Fifth International Conference on Pattern Recognition*, Miami Beach, December 1980.

[8] M. Keil and J. Sack, ''Minimum decompositions of polygonal objects,'' in *Computational Geometry* (ed. G.T. Toussaint), North-Holland, 1985, pp. 197-216.

[9] G.T. Toussaint, ''The relative neighborhood graph of a finite planar set,'' *Pattern Recognition* **12** (1980), pp. 261-268.

[10] R.B. Urquhart, ''Algorithms for computation of relative neighborhood graph,'' *Electronics Letters* **16** (1980), pp. 556-557.

[11] G.T. Toussaint, ''Comment on: Algorithms for computing the relative neighborhood graph,'' *Electronics Letters* **16** (1980), pp. 860-861.

[12] K.J. Supowit, ''The relative neighborhood graph with an application,'' *Journal of ACM* **30** (1983), pp. 428-448.

[13] J. O'Rourke, ''Computing the relative neighbourhood graph in the L_1 and L_∞ metrics,'' *Pattern Recognition* **15** (1982), pp. 189-192.

[14] D.T. Lee, ''Relative neighbourhood graphs in the L_1 metric,'' *Pattern Recognition* **18** (1985), pp. 327-332.

[15] J. Toriwaki and S. Yokoi, ''Voronoi and related neighbours on digitized two-dimensional space with applications to texture analysis,'' in this volume.

[16] J. Katajainen and O. Nevalainen, ''Computing relative neighbourhood graphs in the plane,'' *Pattern Recognition* **19** (1986), pp. 221-228.

[17] G.T. Toussaint and R. Menard, ''Fast algorithms for computing the planar relative neighbourhood graph,'' in *Methods of Operations Research*. Köln, 1980, pp. 425-428.

[18] J.W. Jaromczyk and M. Kowaluk, ''A note on the relative neighbourhood graph,'' *Symposium on Computational Geometry*, Waterloo, Canada, June 1987.

[19] D.W. Matula and R.R. Sokal, ''Properties of Gabriel graphs relevant to geographic variation research and the clustering of points in the plane,'' *Geographical Analysis* **12** (1980), pp. 205-222.

[20] M.I. Shamos, *Problems in Computational Geometry*, Carnegie-Mellon University, 1977.

[21] H. ElGindy and D. Avis, ''A linear algorithm for computing the visibility polygon from a point,'' *J. of Algorithm* **2** (1981), pp. 186-197.

[22] D.T. Lee, ''Visibility of a simple polygon,'' *Computer Vision, Graphics and Image Processing* **22** (1983), pp. 207-221.

[23] G.T. Toussaint, ''A hierarchy of simple polygons,'' manuscript.

Computational Morphology
G.T. Toussaint (editor)

Polygonal Approximations of a Curve — Formulations and Algorithms

Hiroshi IMAI* and Masao IRI†

*Department of Computer Science and Communication Engineering
Kyushu University, Hakozaki, Fukuoka 812, Japan

†Department of Mathematical Engineering and Instrumentation Physics
University of Tokyo, Bunkyo-ku, Tokyo 113, Japan

Abstract

A general framework for polygonal approximation problems in various situations is given, and solution algorithms are discussed. The problems are classified according to three criteria: (1) whether the curve is the graph of a piecewise linear function $y=f(x)$ of the independent variable x, or it is a general polygonal curve in the plane, (2) which error criterion is adopted (for a piecewise linear function, we consider only one criterion), (3) whether the objective is to minimize the number of vertices of the approximate curve when an error bound is given, or to minimize the approximation error when the number of vertices of the approximate curve is specified. The problems are first formulated in terms of graph theory, and then computationally efficient algorithms are constructed by taking advantage of fundamental algorithms for convex hulls in computational geometry.

1. Introduction

Piecewise linear curves are often used to approximate complex boundaries of figures in cartography, computer graphics, pattern recognition, etc. The general problem of approximating a given piecewise linear curve by another coarser one is of fundamental importance. There have been a number of papers for various kinds of polygonal approximation problems: e.g., Kurozumi and Davis [9], and also recent papers by Imai and Iri [7] and Toussaint [16]. Although polygonal approximation problems have been studied for over a decade, it seems that until quite recently no algorithm has been proposed that considers the optimality of the solution and the complexity of the algorithm at the same time. In this paper, we present a unified approach to these problems, and discuss efficient algorithms from the computational-geometric point of view. The presented algorithms are simple enough to run fast not only theoretically but also in practice. The field of polygonal approximation will be one of the most fruitful fields of application of the theory of computational geometry to practical problems.

The polygonal approximation problems examined in this paper are as follows. An approximate curve must satisfy some error criterion, which is specified appropriately for each application. For the approximation of a function, the error of an approximate function is defined as the maximum of the absolute value of the difference between the values of the two functions. In more general cases, a number of criteria have so far been considered, of which four will be considered in this paper.

As is stated in Kurozumi and Davis [9], there are two types of optimization problems connected with polygonal approximation problems, i.e., the problem of

(1) minimizing the number of vertices of an approximate curve with the error within a given bound,

and that of

(2) minimizing the error of an approximate curve consisting of a given number of vertices.

Problems of type (1) will be referred to as min-# problems, and those of type (2) as min-ε problems.

In section 2, we discuss approximation problems for a piecewise linear function. We first outline a linear-time algorithm for the min-# problem given in Imai and Iri [8]. The algorithm employs a simple convex-hull algorithm for monotone polygonal curves, and is efficient in practice. Then we discuss some difficulties that arise in the min-ε problem for a piecewise linear function.

In section 3, we consider approximation problems for general piecewise linear curves (not necessarily functions). We first propose a formulation which unifies approximation problems with different error criteria, together with a graph-theoretic approach to solution algorithms. The approximation problems are reduced to the simple shortest-path problem on graphs associated with the given polygonal curves. In the case of the error criterion proposed in Imai and Iri [7], the corresponding graph has a special structure, which admits more efficient algorithms than in the other cases. In order to improve the efficiencies of the algorithms, fundamental algorithms established in computational geometry are utilized, such as a linear-time algorithm for finding a convex hull of two convex polygons or that of a simple polygon, and the so-called "rotating-caliper" method (see Toussaint [15]).

Finally, the results are summarized and some remarks on extensions of polygonal approximation problems are given.

2. Approximation problems for a piecewise linear function

A function $f: [x^-, x^+] \to \mathbf{R}$ is said to be *continuous piecewise linear* or, simply, *piecewise linear* if its graph $y = f(x)$ in the xy-plane is a polygonal line connecting points p_1, p_2, ..., p_n in this order such that $x^- = x(p_1) < x(p_2) < \ldots < x(p_n) = x^+$, where $x(p)$ denotes the x-coordinate of point p. This polygonal line is denoted by $p_1p_2 \ldots p_n$, and will often be identified with the function f itself. This polygonal line is strictly monotone with respect to the x-axis (hereafter, in this section, "monotone" will always mean "strictly monotone with respect to the x-axis"). Another piecewise linear function $\tilde{f}$ is said to approximate f with error bound ε if the x-coordinates of the two endpoints of the polygonal line of $y=\tilde{f}(x)$ coincide with those of f, i.e., with $x(p_1)$ and $x(p_n)$, and if $|\tilde{f}(x)-f(x)|\leqq\varepsilon$ for any x between $x(p_1)$ and $x(p_n)$.

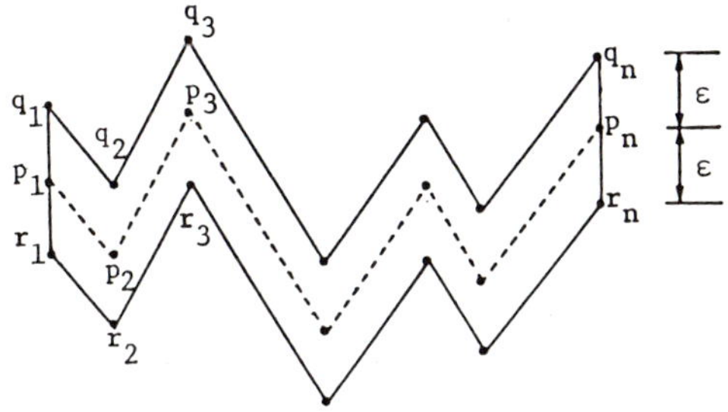

Fig.2.1. Polygonal line $p_1p_2...p_n$ (broken line) and polygon $P(\varepsilon)$ (solid line)

For a polygonal line $p_1p_2...p_n$, a polygon is constructed by sliding it vertically by ε both upwards and downwards. We shall denote this polygon by $P(\varepsilon)$ (see Fig.2.1). We denote the vertices of the upper (resp. lower) boundary of $P(\varepsilon)$ by $q_1, q_2, ..., q_n$ (resp. $r_1, r_2, ..., r_n$) corresponding to $p_1, p_2, ..., p_n$. A polygonal line contained in $P(\varepsilon)$ which connects a point on edge e and a point on edge e' will be called simply a polygonal line connecting edges e and e'. Then,

Observation 2.1. A piecewise linear function $\tilde{f}$ is an approximate piecewise linear function of f with error bound ε if and only if its polygonal line is a monotone polygonal line connecting edge $\overline{q_1r_1}$ and edge $\overline{q_nr_n}$ of polygon $P(\varepsilon)$. □

Thus, the approximation problem for a piecewise linear function is reduced to that of finding a monotone polygonal line from $\overline{q_1r_1}$ to $\overline{q_nr_n}$.

2.1. The min-# approximation problem for a piecewise linear function

Based on Lemma 2.1, let us consider the problem of finding a monotone polygonal line with the minimum number of vertices (points) which connects edge $\overline{q_1r_1}$ and edge $\overline{q_nr_n}$ of polygon $P(\varepsilon)$, where the concept of *edge visibility*, first studied by Avis and Toussaint [2], plays an important role. A point u in a polygon P is said to be visible from an edge e of P if there exists a point v on e such that the line segment $\overline{uv}$ is in P. An edge e' of P is said to be visible from edge e if there is a point on e' which is visible from e. The visibility polygon VP(P,e) from an edge e of polygon P is defined as the set of points in P that are visible from edge e (Fig.2.2). The polygon P is partitioned into several connected regions by the visibility from e, i.e., into VP(P,e) and the remaining polygons consisting of the points invisible from e. The latter polygons will be referred to as *invisible polygons*. When e' is not visible from e, there is an invisible polygon P' one of whose edges is e'. The common boundary of VP(P,e) and the invisible polygon

P' is an edge of P', which is called the *window* (in P) from e to e'. (In Fig.2.2, the window from e to e' is $\overline{uv}$.)

Making use of the visibility concepts let us consider the following algorithm for $P(\varepsilon)$.

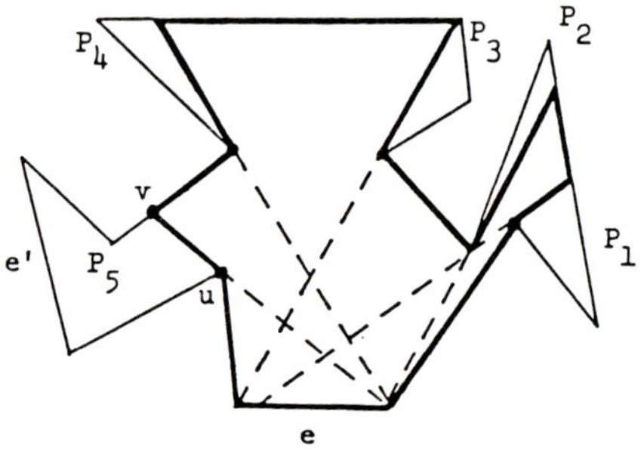

Fig.2.2. Visibility polygon VP(P,e) (thick line) from edge e

Algorithm:

1. $P_1 := P(\varepsilon)$; $e_1 := \overline{q_1 r_1}$; $i := 1$;
2. **while** $\overline{q_n r_n}$ is not visible from e_i in polygon P_i **do**

   begin

   e_{i+1} := the window from e_i to $\overline{q_n r_n}$ in polygon P_i;

   P_{i+1} := the invisible polygon of e_i containing $\overline{q_n r_n}$ in polygon P_i;

   s_i := the point of intersection of edge e_i and the line containing e_{i+1};

   $i := i+1$

   end;
3. $m := i+1$;

   find a point s_{m-1} on edge e_{m-1} and a point s_m on edge $\overline{q_n r_n}$ which are visible from each other in polygon P_{m-1}.

At the end of the algorithm, $s_1 s_2 ... s_m$ is an approximate polygonal line having the minimum number of points. (Intuitively, this may be natural. For the formal proof see Imai and Iri [8].) See also Fig.2.3 to see how the algorithm proceeds.

From the algorithmic viewpoint, it is crucial to find the relevant windows efficiently. So we have to investigate the geometric properties of windows. In polygon $P(\varepsilon)$, let q be a point on $\overline{q_{i-1} q_i}$ (distinct from q_i) and r be a point on $\overline{r_{j-1} r_j}$ (distinct from r_j) such that q and r are visible from each other. We can then consider a simple polygon P of which the upper boundary is $q q_i q_{i+1} ... q_n$ and the lower boundary is $r r_j r_{j+1} ... r_n$, each of them being a monotone polygonal line. Let us consider the window from edge $\overline{qr}$ to $\overline{q_n r_n}$ in this polygon P (see Fig.2.4). Then, we have the following lemma, where we denote the convex hull of a polygonal line $t_1 t_2 ... t_k$ by CH($t_1 t_2 t_3 .... t_k$).

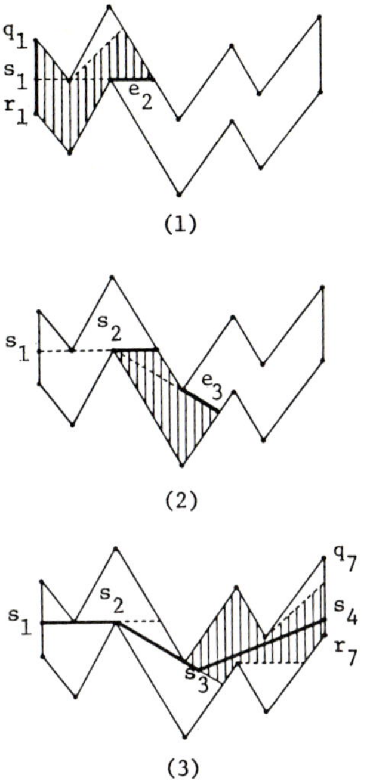

Fig.2.3. Illustration of Algorithm

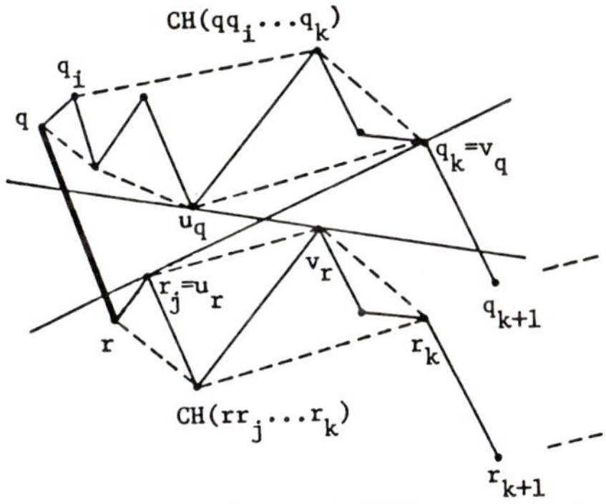

Fig.2.4. The convex hulls (dotted areas), CH($qq_i...q_k$) and CH($rr_j...r_k$), of polygonal lines $qq_i...q_k$ and $rr_j...r_k$, and their separating lines, u_rv_q and u_qv_r

Lemma 2.1. For $k \geqq \max\{i,j\}$, $\overline{q_kr_k}$ is visible from $\overline{qr}$ if and only if CH($qq_iq_{i+1}...q_k$) and CH($rr_jr_{j+1}...r_k$) have no common interior point (but they may touch each other).□

Lemma 2.2. If, for a $k \geqq \max\{i, j\}$, CH($qq_iq_{i+1}...q_k$) and CH($rr_jr_{j+1}...r_k$) have no common interior point, and u_q, u_r, v_q, v_r are the four supporting points of the two separating lines (as illustrated in Fig.2.4), then the following propositions hold.

(i) If q_{k+1} is above the line $u_q v_r$ or on it and r_{k+1} is under the line $u_r v_q$ or on it, then CH($qq_i ... q_{k+1}$) and CH($rr_j ... r_{k+1}$) have no common interior point.

(ii) If q_{k+1} is under the line $u_q v_r$, then CH($qq_i ... q_{k+1}$) and CH($rr_j ... r_{k+1}$) have a common interior point, and the window from $\overline{qr}$ to $\overline{q_n r_n}$ is $\overline{v_r w}$ where w is the point of intersection of line $u_q v_r$ and edge $\overline{q_k q_{k+1}}$.

(iii) If r_{k+1} is above the line $u_r v_q$, then CH($qq_i ... q_{k+1}$) and CH($rr_j ... r_{k+1}$) have a common interior point, and the window from $\overline{qr}$ to $\overline{q_n r_n}$ is $\overline{v_q w}$ where w is the point of intersection of line $u_r v_q$ and edge $\overline{r_k r_{k+1}}$. □

Based on these lemmas, we arrive at the following algorithm for finding the required window: Starting from $q=q_1$, $r=r_1$ and $k=3$ (the case of $k=2$ is trivial), check whether CH($q ... q_k$) and CH($r ... r_k$) intersect for $k=3,4,5,...$; if they have a common interior point, find the window from $\overline{qr}$ to $\overline{q_n r_n}$ by using Lemma 2.2, and updating $\overline{qr}$ to this window, repeat this procedure until $\overline{q_n r_n}$ becomes visible from $\overline{qr}$.

Since $q ... q_k$ and $r ... r_k$ are monotone, their convex hulls can be computed efficiently, and we can prove the following (see Imai and Iri [8] for more details).

Theorem 2.1. The min-# approximation problem for a continuous piecewise linear function with n corner points can be solved in $O(n)$ time. □

2.2. The min-ε approximation problem for a piecewise linear function

Although the min-# problem for a piecewise linear function can be solved in time linear to the number of corner points, the min-ε problem for it is harder to solve.

Consider the problem of finding a minimum-error approximate piecewise linear curve consisting of two segments for the piecewise linear curve depicted in Fig.2.5. It is seen after some calculation that the minimum error is $(21-\sqrt{57})/16$ although the coordinates of the given points are all integer-valued; that is, the rational field is not sufficient but we need its algebraic extension of degree two in this case. In general, in

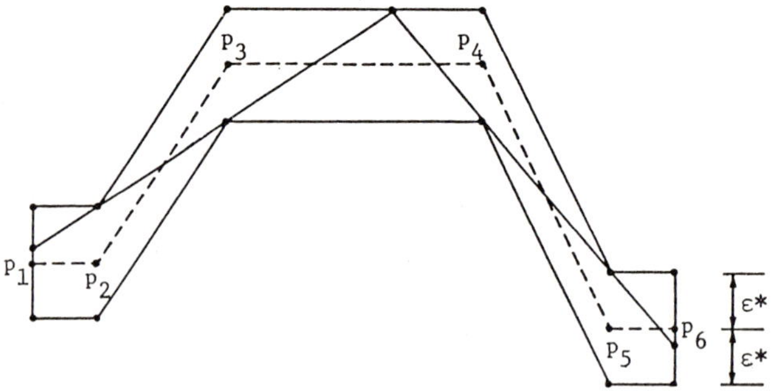

Fig.2.5. An example of the min-ε problem with the minimum error $\varepsilon^*=(21-\sqrt{57})/16$ for $m=2$ ($p_1=(0,0)$, $p_2=(1,0)$, $p_3=(3,3)$, $p_4=(7,3)$, $p_5=(9,-1)$, $p_6=(10,-1)$)

order to obtain the minimum error of an approximate curve of k segments, we need an algebraic extension of degree k, which we fear would make the optimal ε-value hard to determine.

In Imai [6], a special case of the min-ε problem is shown to be solvable in $O(n^2 \log n)$ time.

3. The approximation problems for a general piecewise linear curve

There are quite a number of definitions of "approximation" of a general polygonal curve. Here, in order to treat the problems in a unified way, we adopt the following definition. For a given polygonal curve whose vertices are $p_1, ..., p_n$ in this order, an approximate polygonal curve for it consists of vertices $p_{i(1)}, p_{i(2)}, ..., p_{i(m)}$ which are a subset of the points $p_1, ..., p_n$ of the given curve (where $1 \equiv i(1) < i(2) < ... < i(m) \equiv n$). The error of the approximate curve is defined to be the maximum among the "errors" of its segments $\overline{p_{i(l)}p_{i(l+1)}}$ ($l=1, ..., m-1$). Various types of "errors of segments" have been considered. In this paper, we examine the following four error criteria (see Fig.3.1):

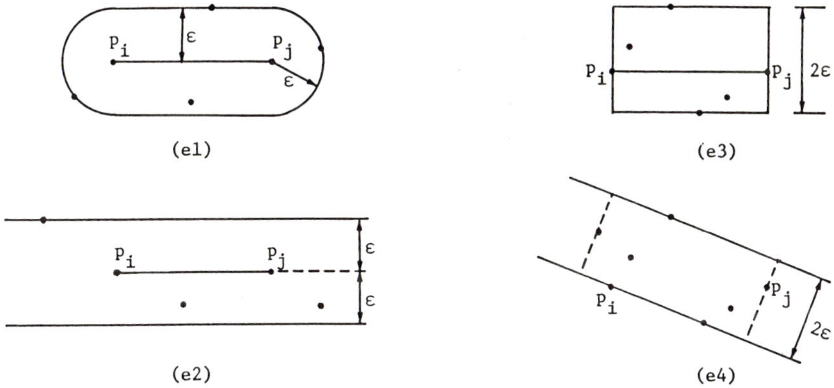

Fig.3.1. Error criteria (e1), (e2), (e3) and (e4)

The error of segment $\overline{p_ip_j}$ ($i<j$) is defined to be

(e1) the maximum of the distance between the segment $\overline{p_ip_j}$ and the points p_k ($i \leqq k \leqq j$);

(e2) the maximum distance between the *line* connecting p_i and p_j and the points p_k ($i \leqq k \leqq j$) (termed the "parallel-strip" criterion in Toussaint [16]);

(e3) the minimum width (divided by 2) of a rectangle containing all the points p_k ($i \leqq k \leqq j$) such that p_i and p_j are on its two opposite sides, which are orthogonal to segment $\overline{p_ip_j}$ (the remaining two opposite sides being parallel to $\overline{p_ip_j}$), where the width of the rectangle is the distance of the two sides parallel to $\overline{p_ip_j}$. The error of $\overline{p_ip_j}$ is $+\infty$ if no such rectangle exists (Ballard [3]; to be named "parallel-rectangle" criterion);

(e4) the minimum width (divided by 2) of the rectangle which contains all the points p_k ($i \leqq k \leqq j$), where the width of the rectangle is the smaller of the distances between the pairs of sides.

The error criteria (e3) and (e4) correspond to covering a sequence of points by rectangles, and so the "approximate curves" for these two criteria might look peculiar. Also, the criterion (e4) is used to find an approximate polygonal curve whose points are not necessarily a subsequence of the points of the original curve; see Imai and Iri [7].

Sometimes a given polygonal curve will be a closed curve (i.e., a circuit) so that a approximate closed curve will be required. Specifically, for a closed polygonal curve whose points are $p_1, p_2, ..., p_n$ in this circular order, its approximate curve is an approximate curve of an "open" polygonal curve connecting $p_i, ..., p_n, p_1, ..., p_{i-1}, p_i$ for some i. The approximation problem of the min-# type as well as that of the min-ε type for a closed curve, which will be referred to as the closed min-# problem and the closed min-ε problem, respectively, are defined in a manner similar to the corresponding problem for an open curve.

3.1. A general approach using graphs

First, let us introduce the associated graph for a given polygonal curve $C=p_1p_2...p_n$. Consider the directed graph, denoted by $G(C)$, having a vertex v_i corresponding to each point p_i of C and an arc (v_i,v_j) for each pair (i, j) such that $1 \leqq i < j \leqq n$, where each arc (v_i,v_j) is assigned a weight which is the error of segment $\overline{p_ip_j}$. For a given error bound ε, define a graph, $G(C,\varepsilon)$, to be the subgraph of $G(C)$ consisting of the arcs whose weights are less than or equal to ε. $G(C)$ and $G(C,\varepsilon)$ are by definition acyclic (since C is not closed), and a path in $G(C,\varepsilon)$ corresponds to an approximate curve with error at most ε.

In the min-# approximation problems, an approximate curve with the minimum number of vertices corresponds to a shortest path from v_1 to v_n in $G(C,\varepsilon)$, where the length of a path is the number of its arcs. The shortest path can be found simply by a breadth-first search in time linear in the number of arcs in $G(C,\varepsilon)$. On the other hand, the min-ε problem is the problem of minimizing ε such that, in $G(C,\varepsilon)$, there is a path from v_1 to v_n consisting of a given number of points. Thus ,we have the following.

Lemma 3.1. (i) If $G(C)$ can be computed in $f(n)$ time, the min-# problem can be solved in $O(f(n)+n^2)$ time and the min-ε problem can be solved in $O(f(n)+n^2 \log n)$ time.

(ii) If $G(C,\varepsilon)$ can be computed in $g(n)$ time, the min-# problem can be solved in $O(g(n)+n^2)$ time.

Proof: $G(C,\varepsilon)$ can be computed in $O(n^2)$ time from $G(C)$, and the shortest path problem on $G(C,\varepsilon)$ can be solved in $O(n^2)$ time by a breadth-first search. Hence the propositions for the min-# problems hold. For the min-ε problem with m points, the minimum ε^* should be equal to the weight of an arc in $G(C)$, and for any ε greater than ε^*, there is a path from v_1 to v_n consisting of m points in $G(C,\varepsilon)$ and for any ε smaller than ε^* there is no such path in $G(C,\varepsilon)$. Hence, ε^* can be found by first sorting the $n(n-1)$ weights of the arcs in $G(C)$ and then executing a binary search. Sorting can be done in $O(n^2 \log n)$ time. The binary search will have $O(\log n)$ steps, each requiring the solution of a shortest path problem, solvable in $O(n^2)$ time. □

Concerning the closed approximation problems, we trivially have the following.

Lemma 3.2. The closed min-# (resp. min-ε) problem can be solved by solving the n ordinary "open" min-# (resp. min-ε) problems. □

The closed approximation problems under error criterion (e4) can be solved more efficiently than indicated in Lemma 3.2 owing to the special structure of the associated graphs, which will be discussed in section 3.2.

Toussaint [16] shows that, by using the on-line convex hull algorithms (e.g., Avis, ElGindy and Seidel [1] and Preparata [11]), the graph $G(C)$ under error criterion (e2) can be constructed in $O(n^2 \log n)$ time and hence the min-# problem under (e2) can be solved in $O(n^2 \log n)$ time.

It can be seen without difficulty that graph $G(C)$ under error criterion (e3) can also be constructed in $O(n^2 \log n)$ time by means of the on-line convex hull algorithms, as is shown in Toussaint [16].

Theorem 3.1. Under error criterion (e2) and (e3), both the min-ε problem and the min-# problem can be solved in $O(n^2 \log n)$ time. □

Melkman and O'Rourke [10] have shown that $G(C,\varepsilon)$ under error criterion (e1) can be constructed in $O(n^2 \log n)$ time by essentially extending the so-called cone-intersection method [13], [17] and hence the min-# problem under (e1) can be solved in $O(n^2 \log n)$ time. However, it seems that their algorithm cannot be applied to the construction of $G(C)$. $G(C)$ itself can be constructed in $O(n^3)$ time by a naive algorithm of computing the error of each segment $\overline{p_i p_j}$ in $O(n)$ time, from which it is readily seen that the min-ε problem under (e1) can be solved in $O(n^3)$ time. However, there is a much better algorithm which partly makes use of Melkman and O'Rourke's algorithm.

Lemma 3.3. The minimum error ε^* under (e1) is contained in a set $E(C)$ defined by

$$E(C) = \{\text{distance of } p_i \text{ and } p_j \mid 1 \leqq i < j \leqq n\}$$
$$\cup \{\text{error of segment } \overline{p_i p_j} \text{ under the criterion (e2)} \mid 1 \leqq i < j \leqq n\}. \quad \square$$

Since $|E(C)|$ is $O(n^2)$, we can determine the set $E(C)$ in $O(n^2 \log n)$ time by means of the method of Toussaint [16] . Then, by executing binary search on $E(C)$ to find ε^* (where each step of binary search can be done in $O(n^2 \log n)$ time using Melkman and O'Rourke's algorithm [10]), we can find ε^* in $O(n^2(\log n)^2)$ time. Thus, we have

Theorem 3.2. The min-ε problem under error criterion (e1) can be solved in $O(n^2(\log n)^2)$ time. □

Under error criterion (e4), we can construct $G(C)$ in $O(n^3)$ time by means of the on-line convex-hull algorithms and the rotating-caliper method, but the approximation problems under (e4) may be solved in a more efficient way than by constructing $G(C)$, $G(C,\varepsilon)$ explicitly, which is discussed next.

3.2. The approximation problems under (e4)

Under the error criterion (e4), $G(C,\varepsilon)$ has the following nice property.

Lemma 3.4. In $G(C,\varepsilon)$, if there is an arc (v_i,v_j), then there are arcs (v_k,v_l) for all pairs (k,l) with $i \leqq k < l \leqq j$. □

Hence, the shortest path $v_1 = v_{i(1)}, v_{i(2)}, \ldots, v_{i(\mu)} = v_n$ on $G(C,\varepsilon)$ can be found by a simple greedy algorithm:

Algorithm G0:

1. $\nu := 1$; $i(1) := 1$;
2. $i(\nu+1) := \max\{j \mid \text{arcs } (v_{i(\nu)}, v_j) \text{ in } G(C,\varepsilon)\}$;
3. **if** $i(\nu+1) = n$ **then** halt **else** $\nu := \nu + 1$ and return to 2.

Under (e4), this construction of $G(C,\varepsilon)$ is still rather costly (it takes $O(n^2 \log n)$ time). In the following, we describe the algorithms in Imai and Iri [7] which make full use of the property stated in Lemma 3.4. From here on, we shall view the approximation problems as covering a sequence of points by rectangles as we did in [7], since this is a more intuitive way to describe and understand the algorithms.

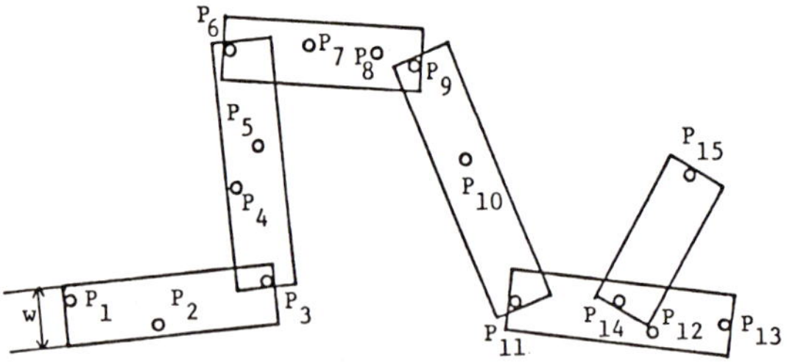

Fig.3.2. Covering a sequence of points by rectangles

Given a sequence of n points $p_1, p_2, \ldots, p_n$, we say that rectangles $R_1, R_2, \ldots, R_r$ (in sequence) with width $w(=2\varepsilon)$ cover the sequence of points if there are indices $i(0)$, $i(1), \ldots, i(r)$ $(1 \equiv i(0) < i(1) < \ldots < i(r-1) < i(r) \equiv n)$ such that, for each $q = 1, \ldots, r$, rectangle R_q contains points $p_{i(q-1)}, \ldots, p_{i(q)}$. By a rectangle with width w, we mean a rectangle with width w and a sufficiently large length. When we consider a covering rectangle with width w, we suppose that it has the minimal length so long as it covers what it should. Then the min-# problem is the problem of finding the minimum number of rectangles (in sequence) of width w which cover the given sequence of points (see Fig.3.2). The min-ε problem is the problem of minimizing the width w such that the given sequence of points can be covered by at most m rectangles with width w. We shall give an $O(n \log n)$-time algorithm for the former problem, and an $O(mn (\log n)(\log (n/m)))$-time algorithm for the latter.

3.2.1. The min-# problem under (e4)

First, let us consider the problem of determining whether or not a set of points can be covered by a rectangle with width w. For this problem, the following lemma is almost obvious (see, e.g., [5]).

Lemma 3.5. There is a rectangle with width w which can cover a set of points if and only if there is a rectangle with width w covering the set of points such that an edge of the convex hull of those points lies on one of the two parallel sides of distance w of the rectangle (Fig.3.3). □

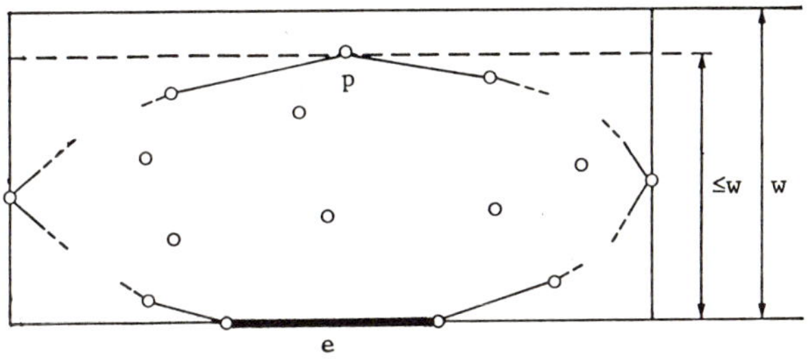

Fig.3.3. The convex hull of points and a covering rectangle (p is the antipodal point of e)

Thus, the problem is directly connected with the efficient construction of the convex hulls of points. For points $p_1, ..., p_n$, we denote by $CH(i,j)$ the convex hull of points $p_i, ..., p_j$ $(1 \leqq i \leqq j \leqq n)$. We also denote by $W(i,j)$ the minimum width of a rectangle that can cover points $p_i, ..., p_j$. For an edge e of a convex hull, we call the vertex of the convex hull that is the farthest from the straight line containing e the "*antipodal point*" of e, and denote by $d(e)$ the distance of the antipodal point of e from the straight line containing e. Then, from Lemma 3.5, we have

$$W(i,j) = \min\{ d(e) \mid e \text{ is on } CH(i,j) \}.$$

Given a convex hull with N edges, we can find the antipodal points of all the edges in $O(N)$ time by means of the so-called rotating-caliper method (Shamos [12], Toussaint [15]), well known in computational geometry.

The convex hull of n points can be found in $O(n \log n)$ time (e.g., Graham [4]), so that we have the following lemma (a similar lemma for the Chebyshev approximation of a set of points by a line is shown in Shamos [12]).

Lemma 3.6. $W(1,n)$ can be computed in $O(n \log n)$ time. Furthermore, when the convex hull CH(1,n) is given, $W(1,n)$ can be computed in $O(n)$ time. □

We now consider the problem of finding the minimum number of rectangles with width w that cover a sequence of points $p_1, \ldots, p_n$. The greedy algorithm for the shortest path problem in $G(C,\varepsilon)$ can be restated in the case of rectangular covering as follows.

Algorithm G (Greedy Algorithm):

G1. $\nu := 1$; $i(1) := 1$;

G2. Find $i(\nu+1) := \max\{j \mid i(\nu) < j \leqq n,\ W(i(\nu), j) \leqq w\}$;

G3. Cover points $p_i, \ldots, p_{i(\nu)}$ by a rectangle with width w;
if $i(\nu+1) = n$ **then** halt **else** $\nu := \nu+1$ and return to G2.

The main step of Algorithm G is step G2. We shall show that this step G2 for $\nu = 1$ can be executed in $O(i^* \log i^*)$ time where $i^* = i(2)$. Since $W(1,j)$ is nondecreasing in j, we can employ a doubling binary search technique to find i^*. First, we find m and l such that $W(1,m) \leqq w$, $W(1,l) > W$ and $m < l \leqq \min\{2m, n\}$ (or we find $W(1,n) \leqq w$) by a doubling search; i.e., starting with $m=1$ and $l=2$, we repeat doubling both m and l until the condition is satisfied. Then, i^* is known to lie in the interval between m and $l-1$. The exact value of i^* can be determined by means of binary search by the following algorithm, whose the validity is obvious.

Algorithm B (Binary search):

B1. $k := m$; $l := l-1$; Construct and retain CH(1,k);

B2. $h := \lceil (k+l)/2 \rceil$;
Construct CH($k+1$,h);
Construct CH(1,h) from CH(1,k) and CH($k+1$,h);
Compute $w_h := W(1,h)$ by the rotating-caliper method;

B3. **if** $w_h \leqq w$ **then** $k := h$ **else** $l := h-1$;
if $k = l$ **then** $i^* := k$ and halt **else** retain CH(1,k) and return to B2.

The complexity of this Algorithm B is evaluated as follows. The execution of the doubling search is dominated by the computation of $W(1,2^g)$ for $g = 1, 2, \ldots, M+1$ where $M = \log_2 m$. Since $W(1,j)$ can be found in $O(j \log j)$ time (Lemma 3.6), we can execute the doubling search in $O(\Sigma_{g=1}^{M+1} 2^g \log 2^g) = O(i^* \log i^*)$ time (note that $m = O(i^*)$).

In Algorithm B, all the steps except the part of B2 for finding CH($k+1$,h) can be executed in $O(i^* \log i^*)$ time, because, in B2, CH(1,h) can be found in $O(h)$ time by computing the convex hull of two convex polygons CH(1,k) and CH($k+1,h$) (Shamos [12]), and then $W(1,h)$ can be computed in O(h) time by the rotating-caliper method. In the gth execution of B2, CH($k+1,h$) is the convex hull of at most $m2^{-g}=2^{M-g}$ points, which can be found in $O(2^{M-g}\log 2^{M-g})$ time. Since B2 is executed at most M times, computation of CH($k+1,h$) in the whole execution of Algorithm B takes $O(\Sigma_{g=1}^{M} 2^{M-g}\log 2^{M-g})=O(i^*\log i^*)$ time.

In Algorithm G, step G2 for general $i(\nu+1)$ can be executed in $O((i(\nu+1)-i(\nu))\log(i(\nu+1)-i(\nu)))$ time by means of the above algorithm, and hence Algorithm G is completed in $O(n \log n)$ time.

3.2.2. The min-ε problem under (e4)

We now consider the problem of minimizing the width w such that a sequence of n points can be covered by at most m rectangles with width w, and present an $O(mn(\log n)(\log(n/m)))$-time algorithm. The algorithm is simple, and is described as follows, where we set $W(i,n+1)=\infty$ $(1\leqq i\leqq n)$.

Algorithm A (Compute the minimum width w^*):

A1. $m':=m$; $i:=1$; $w:=0$;

A2. Find j, by a doubling binary search, such that
- (i) $i<j\leqq n$,
- (ii) the sequence of points $p_i, ..., p_n$ cannot be covered by m' or fewer rectangles with width less than $W(i,j)$;
- (iii) the sequence of points $p_i, ..., p_n$ can be covered by m' or fewer rectangles with width less than $W(i,j+1)$;

A3. $i:=j$; $m':=m-1$; $w:=\max\{w,W(i,j)\}$;
if $i=n$ **then** halt **else** return to A2.

At the end of Algorithm A, w is the minimum width w^*.

Lemma 3.7. Algorithm A correctly computes the minimum width w^*.

Proof: We have only to show the following claim: In a sequence of rectangles with width w^* covering the sequence of points, which is found by Algorithm G, the first rectangle covers points $p_1,..., p_{j_1}$, where j_1 is defined to be that j which is found in the first execution of step A2. (If this is true, then the lemma can be easily proven by induction.) Since a sequence of points $p_1, ..., p_n$ can be covered by at most m rectangles with width less than $W(1,j_1+1)$, we have $w^*<W(1,j_1+1)$. Furthermore, since the sequence cannot be covered by m or fewer rectangles with width less than $W(1,j_1)$, we have $W(1,j_1)\leqq w^*$. Hence, $W(1,j_1)\leqq w^*<W(1,j_1+1)$. This shows that the claim is true. □

As for the time complexity of Algorithm A, step A2 can be executed in $O(n(\log n)^2)$ time by means of the binary search and Algorithm G. Since steps A2 and A3 are executed at most m times, the total time complexity of Algorithm A is $O(mn(\log n)^2)$

time. By more detailed analysis we can show that the factor $(\log n)^2$ in the complexity function can be reduced to $\log n \cdot \log(n/m)$.

3.2.3. The closed min-ε problem under (e4)

Consider a polygon whose vertices (or, points) are given in the clockwise order: $p_1, p_2, \ldots, p_n$. We say that rectangles (in sequence) cover the points of the polygon if they cover a sequence of points $p_i, p_{i+1}, \ldots, p_n, p_1, \ldots, p_i$ for some i with $1 < i \leqq n$.

We consider the problem of minimizing the width w such that the points of the polygon can be covered by at most m rectangles with width w. The minimum width w^* can be found by the following algorithm.

Algorithm:

1. For each i with $1 \leqq i \leqq n$, repeat the following (1.1) and (1.2) for the sequence of points $p_i, \ldots, p_n, p_1, \ldots, p_i \equiv q_0, \ldots, q_n$ ($W(0,j)$ is the minimum width of a rectangle that can cover points $q_0, \ldots, q_j$);

   (1.1) Find j, by a binary search, such that

   - (i) $0 < j \leqq n$,
   - (ii) the sequence of points $q_0, \ldots, q_n$ cannot be covered by m or fewer rectangles with width less than $W(0,j)$;
   - (iii) the sequence of points $q_0, \ldots, q_n$ can be covered by m or fewer rectangles with width less than $W(0,j+1)$ (we set $W(0,n+1) = \infty$);

   (1.2) **if** the sequence of points $q_0, \ldots, q_n$ can be covered by at most m rectangles with width $W(0,j)$ **then** $w_i := W(0,j)$ **else** $w_i := \infty$;

2. Let $w^* = \min\{ w_i \mid 1 \leqq i \leqq n\}$.

The validity of this algorithm can be shown by the arguments similar to the proof of Lemma 3.7. Since step (1.1) for each i can be executed in $O(n(\log n)^2)$ time by means of the binary search and Algorithm G, the complexity of this algorithm is $O((n \log n)^2)$.

4. Concluding Remarks

In this paper, we have formulated general piecewise linear curve approximation problems in a unified framework, and presented efficient algorithms for them. The results are summarized in Table 4.1.

Concerning the approximation problems for general polygonal curves, an approximate curve under error criterion (e1) consists of a subsequence of points of the given curve, whereas error criterion (e4) leads to a non-subsequence approximate curve. Comparing the two from the viewpoint of map size reduction in cartography, for example, e4 will be more useful in practice because it is computationally less complex, i.e., is of order $O(n \log n)$, and, at the same time, e4 ordinarily gives a better approximation in the sense that fewer segments are needed for the same tolerance.

Table 4.1. Summary of the time complexities of polygonal approximation algorithms (n: the number of points of the given polygonal curve).

		min-# problem	min-ε problem
piecewise linear function		$O(n)$	–*
polygonal curve			
	(e1)	$O(n^2 \log n)$**	$O(n^2 (\log n)^2)$
error	(e2)	$O(n^2 \log n)$†	$O(n^2 \log n)$
criteria	(e3)	$O(n^2 \log n)$	$O(n^2 \log n)$
	(e4)	$O(n \log n)$	$O(mn(\log n)(\log (n/m)))$††

* apparently difficult; some special case is solved in $O(n^2 \log n)$ time (Imai [6]), **Melkman and O'Rourke [10], †Toussaint [16], ††m: the (given) number of points of the approximate curve.

Error criteria (e3) and (e4) lead to the problem of covering a sequence of points by rectangles. Such a covering (specifically, the covering under (e3)) is used by Ballard [3] in order to represent a given polygonal curve hierarchically by a data structure called "*strip trees*". Algorithmically, the rectangular covering under (e4) is easier to handle than that under (e3), and hence, another type of "strip tree" by using the rectangular covering under (e4) would deserve investigation, as is pointed out in Imai and Iri [7].

For the general approximation problem, there still remain many problems left open. For example, a challenging problem is to find a simple approximate curve when the given piecewise linear curve is simple (a piecewise linear curve is *simple* if no nonadjacent segments intersect). Also, from the computational-geometric point of view, it would be interesting to apply computational-geometric shortest-path algorithms like the algorithm noted in section 2 and a general algorithm recently proposed by Suri [14] to the polygonal approximation problem, since the algorithm in section 2 using the visibility concept for the simpler class of problems of approximating a piecewise linear function may be regarded as a kind of shortest-path algorithm.

Acknowledgment

The authors would like to thank Professor G. T. Toussaint of McGill University for encourage them to write this paper and for fruitful discussions on the problems. They are also indebted to Professor J. O'Rourke of Johns Hopkins University for valuable comments and information. The work of the first author was supported in part by Research Aid of Inoue Foundation for Science.

References

[1] D. Avis, H. ElGindy and R. Seidel: Simple On-Line Algorithms for Convex Polygons. In "*Computational Geometry*" (G. T. Toussaint, ed.), North-Holland, Amsterdam, 1985, pp.23 – 42.

[2] D. Avis and G. T. Toussaint: An Optimal Algorithm for Determining the Visibility of a Polygon from an Edge. *IEEE Transactions on Computers*, Vol.C-30 (1981), pp.910–914.

[3] D. H. Ballard: Strip Trees: A Hierarchical Representation for Curves. *Communications of the ACM*, Vol.24, No.5 (1981), pp.310 – 321.

[4] R. L. Graham: An Efficient Algorithm for Determining the Convex Hull of a Finite Planar Set. *Information Processing Letters*, Vol.1 (1972), pp.132 – 133.

[5] K. Ichida and T. Kiyono: Segmentation of Plane Curves (in Japanese). *Transactions of the Electronics and Communication Engineers of Japan*, Vol.58-D (1975), pp.689 – 696.

[6] H. Imai: Studies on Algorithms in Computational Geometry (in Japanese). Doctoral Dissertation, University of Tokyo, 1986.

[7] H. Imai and M. Iri: Computational-Geometric Methods for Polygonal Approximations of a Curve. *Computer Vision, Graphics, and Image Processing*, Vol.36 (1986), pp.31 – 41.

[8] H. Imai and M. Iri: An Optimal Algorithm for Approximating a Piecewise Linear Function. *Journal of Information Processing*, Vol.9, No.3 (1986), pp.159 – 162.

[9] Y. Kurozumi and W. A. Davis: Polygonal Approximation by the Minimax Method. *Computer Vision, Graphics, and Image Processing,* Vol.19 (1982), pp.248 – 264.

[10] A. Melkman and J. O'Rourke: On Polygonal Chain Approximation. This book.

[11] F. P. Preparata: An Optimal Real Time Algorithm for Planar Convex Hulls. *Communications of the ACM*, Vol.22 (1979), pp.443 – 450.

[12] M. I. Shamos: Computational Geometry. Ph.D. Thesis, Department of Computer Science, Yale University, 1978.

[13] J. Sklansky and V. Gonzalez: Fast Polygonal Approximation of Digitized Curves. *Pattern Recognition*, Vol.12 (1980), pp.327 – 331.

[14] S. Suri: A Linear Time Algorithm for Minimum Link Paths inside a Simple Polygon. *Computer Vision, Graphics, Image Processing*, Vol.35 (1986), pp.99 – 110.

[15] G. T. Toussaint: Solving Geometric Problems with the "Rotating Calipers". *Proceedings of IEEE MELECON*, Athens, Greece, 1983.

[16] G. T. Toussaint: On the Complexity of Approximating Polygonal Curves in the Plane. *Proceedings IASTED, International Symposium on Robotics and Automation*, Lugano, Switzerland, 1985.

[17] C. M. Williams: An Efficient Algorithm for the Piecewise Linear Approximation of Planar Curves. *Computer Vision, Graphics, and Image Processing*, Vol.8 (1978), pp.286 – 293.

Computational Morphology
G.T. Toussaint (editor)

ON POLYGONAL CHAIN APPROXIMATION

Avraham MELKMAN †
Joseph O'ROURKE §

† Department of Math and Computer Science
Ben-Gurion University of the Negev
P.O. Box 635
Beer-Sheva 84 105, Israel

§ Department of Computer Science
Johns Hopkins University
Baltimore, MD 21218, USA *

Imai and Iri recently described a clever algorithm for approximating a polygonal chain within a given tolerance. Their algorithm requires $O(n^3)$ time in the worst case. In this note it is shown that their algorithm can be improved to $O(n^2 \log n)$ by exploiting the geometrical constraints of the problem.

1. INTRODUCTION

Let $P_1, P_2, \ldots, P_n$ be successive points of an arbitrary (perhaps self-crossing) polygonal chain **C** in the plane. Imai and Iri [2] considered the problem of finding the best approximating subchain $P_{i_1}, P_{i_2}, \ldots, P_{i_m}$, where $i_1 = 1 < i_2 < \cdots < i_m = n$, and m is minimal such that **C** never deviates more than a given tolerance w from the subchain. This latter condition may be more precisely stated as follows. Let L_{ij} be the line segment whose endpoints are P_i and P_j. Then if i and j are two successive indices in the subchain, the chain $P_i, P_{i+1}, \ldots, P_{j-1}, P_j$ remains within w of L_{ij}.

Imai and Iri construct a directed graph G whose nodes correspond to the vertices of **C**, and such that the nodes for vertices P_i and P_j are connected by an arc of G iff **C** remains within w of L_{ij} from P_i to P_j. If each edge is assigned unit weight, then the shortest path from 1 to n yields the approximating chain with the minimum number of segments. Since G is acyclic, this path can be found in time proportional to the number of edges of G, namely in $O(n^2)$.

* The research of this author was partially supported by grants from NSF (DSR 83-51468) General Motors, IBM, and Martin Marietta.

To construct G, they suggest a brute-force check: for each i, and for each $j > i$, check that each edge of the path from P_i to P_j remains within w of L_{ij}. This requires $O(n^3)$ time. In this note we reduce this construction time to $O(n^2 \log n)$, thereby achieving an $O(n^2 \log n)$ approximation algorithm.

Toussaint also improved the Imai and Iri algorithm from $O(n^3)$ to $O(n^2 \log n)$, but for a different error criterion [4]. Four different error criteria are considered in [3]. Here we only discuss the one tolerance measure defined above.

2. THE ALGORITHM

We first observe that if the vertices of C are within tolerance, then the edges between the vertices are within tolerance.

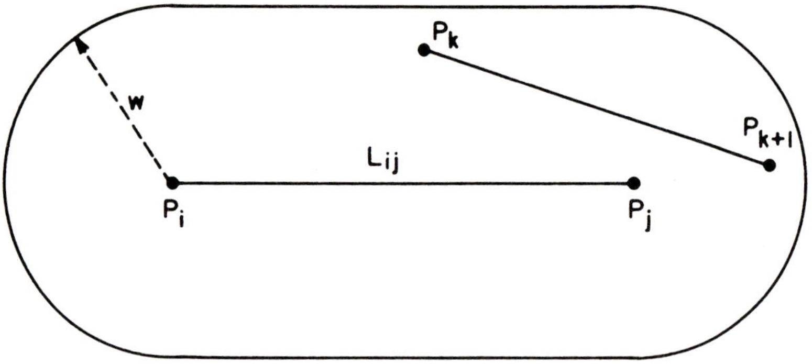

FIGURE 1

Points within w of L_{ij}.

LEMMA 1. If the vertices $P_i, \ldots, P_j$ are each within w of L_{ij}, then the entire path from P_i to P_j is within w of L_{ij}.

Proof. The locus of points that are within a distance w of L_{ij} is bounded by two line segments parallel to L_{ij}, and two semicircles of radius w centered on P_i and P_j; see Fig. 1. Suppose both P_k and P_{k+1}, $i \leq k \leq j$, are each within w of L_{ij}. Then both fall within this oval region. Since the region is convex, all points on the edge $P_k P_{k+1}$ are also in the region, and therefore within the w tolerance.

Thus we need only ensure that the vertices of **C** be within tolerance.

Next we observe that P_k is within tolerance of L_{ij} iff L_{ij} pierces a circle of radius w centered on P_k; call the closed disk bound by this circle C_k. Thus $(i,j) \in G$ iff L_{ij} intersects each C_k, $i \leq k \leq j$.

We can now sketch the algorithm. Index i will be fixed, and j will vary from $i+1$ to n. At each step, arc (i,j) will be output if appropriate. The decision is based upon circle piercing. A "wedge" data structure W_{ij} will be maintained that permits this decision to be reached in $O(\log n)$ time per step, and such that the data structure can be updated in $O(\log n)$ time per step.

Let s be the first index larger than i such that $P_i \notin C_s$. It should be clear that all the arcs $(i,i+1),...,(i,s-1)$ are in G, as all the circles C_j, $i \leq j \leq s$ contain P_i and so are necessarily pierced by L_{ij}. The wedge data structure is not initiated until this first index s whose circle does not include P_i is reached. Moreover, the initial $s-i-1$ vertices of **C** from P_i have no effect on subsequent decisions, and so will be ignored.

A *wedge* W_{ij} is an unbounded wedge-shaped region of the plane with the property that any line segment L_{it} from P_i to a vertex $P_t \in W_{ij}$ necessarily pierces each C_k, $i \leq k \leq j$. To characterize the geometric structure of such a wedge, we need some notation. For a given k, let α_{ik} and β_{ik} be the lower and upper half-infinite rays starting tangent to C_k, aligned with P_i, and going to infinity away from P_i. Denote by D_{ik} the region bound by α_{ik}, β_{ik}, and the arc of C_k, connecting α_{ik} to β_{ik}, that is closer to P_i. These definitions are illustrated in Fig. 2.

LEMMA 2. $W_{ij} = \bigcap_{k=i+1}^{j} D_{ik}$.

Proof. If $P_t \in \bigcap_{k=i+1}^{j} D_{ik}$, then L_{it} must pierce the arc bounding each D_{ik}, and therefore C_k. If $P_t \in W_{ij}$, then by definition it must pierce each C_k, and therefore P_t must lie in each D_{ik}.

It follows from this lemma that the boundary of wedge W_{ij} is composed of two parts: lower and upper boundary rays α and β, and the "frontier" F, which is a convex chain of circular arcs, whose interiors subtend distinct angle ranges with respect to P_i. See Fig. 3. Since all the circles C_k are of the same radius, each can contribute at most one arc to the frontier, for a total of $O(n)$ arcs.

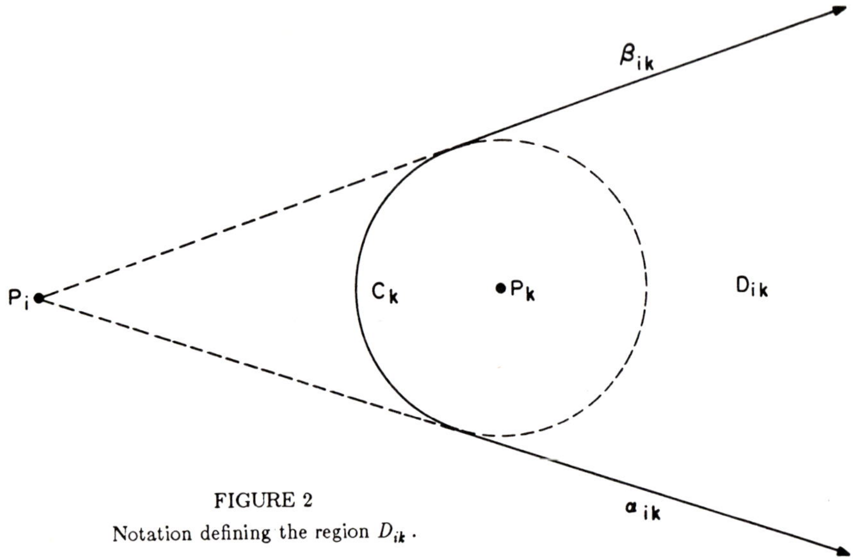

FIGURE 2
Notation defining the region D_{ik}.

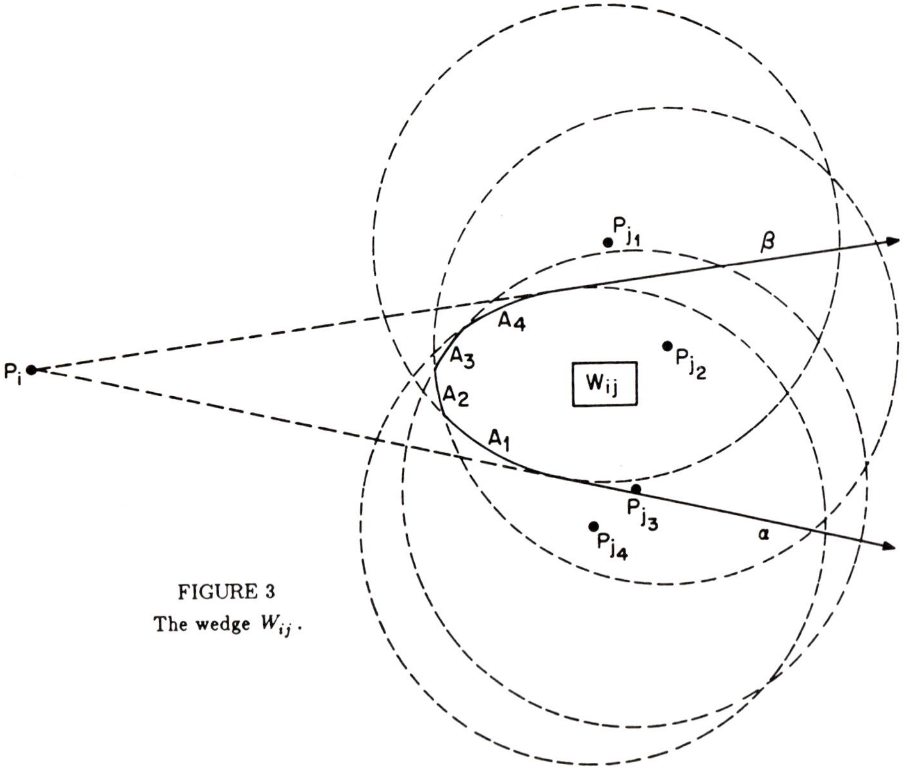

FIGURE 3
The wedge W_{ij}.

The structure established in Lemma 2 is sufficient to achieve an $O(\log n)$ decision time per arc of G. Assume that $W_{i,j-1}$ has been constructed, and that the arc (i,j) is under consideration. If P_j lies strictly in the exterior of $W_{i,j-1}$, then L_{ij} does not pierce each C_k, $i \le k \le j$, and therefore (i,j) is not an arc of G. If $P_j \in W_{i,j-1}$, then L_{ij} pierces each circle and $(i,j) \in G$. Therefore the decision can be made by locating P_j with respect to the angular sectors determined by each arc of the frontier, and then determining on which side of the relevant frontier arc it lies. Since there are $O(n)$ frontier arcs, the location of P_j can be accomplished in $O(\log n)$ time by binary search.

We now turn to the problem of updating wedge $W_{i,j-1}$ to W_{ij}. The updating is accomplished in two stages: first the bounding rays are updated, and next the frontier is modified.

Clearly any ray that pierces C_j must fall within the cone defined by the two tangents to C_j that pass through P_i. This cone is intersected with that defined by the upper and lower bounding edges of $W_{i,j-1}$ to determine new bounding rays α and β. Any portion of the frontier that falls clockwise of α or counterclockwise of β with respect to P_i is clipped in time proportional to the logarithm of the number of arcs in the frontier.

Now we consider updating the frontier. If P_j falls outside $W_{i,j-1}$ but between α and β, then the frontier need not be updated, as any segment that reaches the frontier necessarily pierces C_j. If P_j falls inside the frontier, an arc of C_j may need to be inserted into the frontier, and several arcs deleted.

Let the frontier consist of the arcs $A_1, \ldots, A_r$ in counterclockwise order. It will be convenient to denote $A_0 = \alpha_{i,j-1}$ and $A_{r+1} = \beta_{i,j-1}$.

LEMMA 3. Let A_a and A_b be two intersecting arcs, with respective centers P_a and P_b, such that A_b is counterclockwise from A_a with respect to P_i. Then P_b is clockwise from P_a with respect to P_i.

Proof. Let x be the point of intersection of the two arcs. Then x and P_i are on the same side of the line through P_a and P_b; see Fig. 4. Thus the relative order of the arcs, and their centers, is the same with respect to P_i and x. Since the arcs are part of the boundary of the intersection $C_a \cap C_b$, the lemma follows.

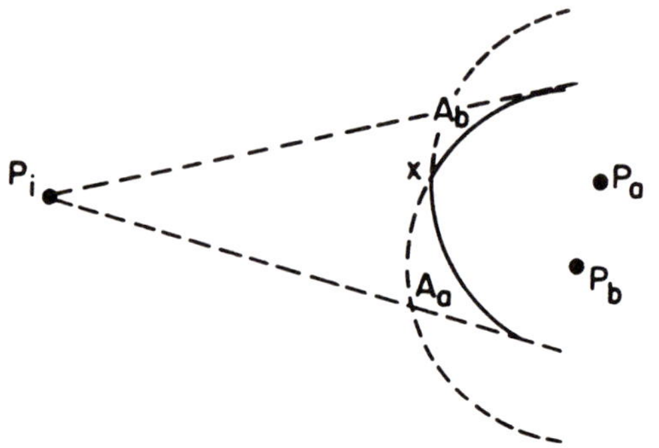

FIGURE 4
Order of centers is reverse of order of arcs.

Corollary. The centers of all the frontier arcs are reverse sorted by angle from P_i.

Thus if the frontier arcs are $A_1, \ldots, A_r$ in counterclockwise order from P_i, and if the arc A_k derives from circle C_{j_k}, then their centers occur in counterclockwise order $P_{j_r}, \ldots, P_{j_1}$. This is illustrated for in Fig. 3 where $r=4$. Our first step is to locate P_j in this list of arc centers in $O(\log n)$ time by binary search. Assume that P_j is found to lie between $P_{j_{k+1}}$ and P_{j_k}. For the next steps we shift attention to the counterclockwise ordered set $\{e_0, \ldots, e_r\}$ of endpoints of the arcs, with e_{i-1} and e_i the endpoints of A_i. First we determine whether any updating is required.

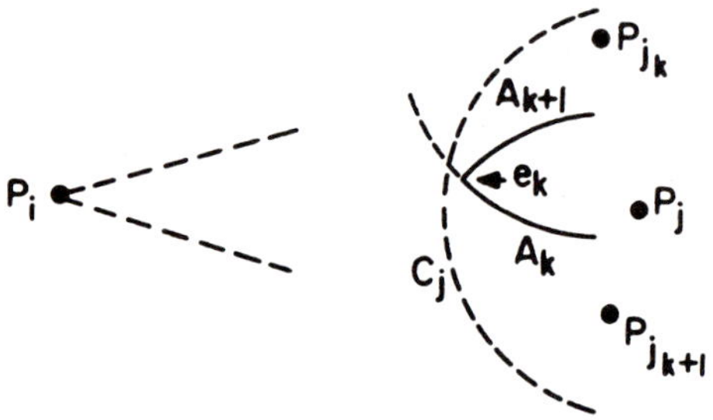

FIGURE 5
No update is necessary at step j.

LEMMA 4. Suppose P_j lies counterclockwise between $P_{j_{k+1}}$ and P_{j_k}, whose corresponding arcs are A_{k+1} and A_k. If $e_k \in C_j$, then $D_{ij} \supset W_{i,j-1}$, i.e., no update is necessary. See Fig. 5.

Proof. Suppose by contradiction that there exists a point z, $z \notin D_{ij}$, but $z \in W_{i,j-1}$. Since $D_{ik} \cap D_{i,k+1} \supset W_{i,j-1}$, $(j > k+1)$, it is also the case that $z \in D_{ik} \cap D_{i,k+1}$. It follows from $e_k \in C_j$ that D_j and $D_{ik} \cap D_{i,k+1}$ intersect but neither includes the other. Therefore C_j must intersect at least on of the arcs A_k, A_{k+1}; say it intersects A_k at x. Then, in the intersection $C_j \cap C_k$, the arc A_k is counterclockwise from the arc corresponding to C_j because $e_k \in C_j$ implies that C_j crosses from being further from P_i than A_k to being closer, when moving counterclockwise through x. Thus, by Lemma 3, P_j must be clockwise from P_{j_k} with respect to P_i, a contradiction.

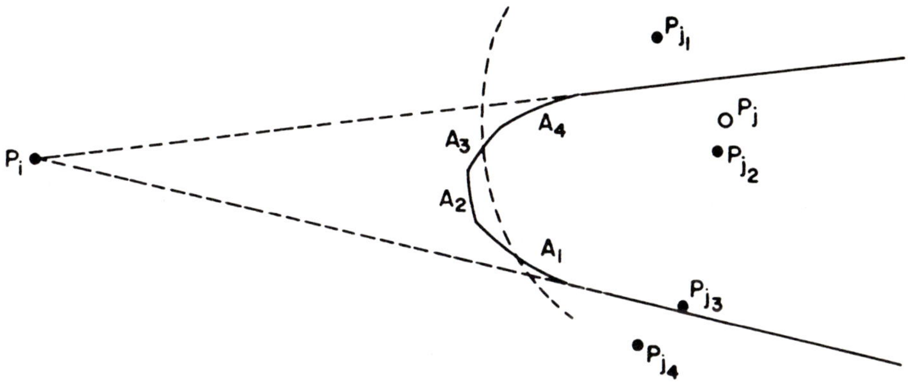

FIGURE 6
Updating the frontier of arcs.

Assume an update is indicated. Observe that C_j may intersect F in at most two locations due to the convexity of F established in Lemma 2. Suppose it intersects arcs A_a and A_b with $0 \le a < b \le r+1$. Then the new frontier for W_{ij} is $A_1, \ldots, A_{a-1}, A_a', A, A_b', A_{b+1}, \ldots, A_r$, where A_a' and A_b' are the portions of A_a and A_b inside C_j, and A is the new arc derived from C_j. The fact that P_j is found to lie between $P_{j_{k+1}}$ and P_{j_k} does *not* imply that C_j intersects arcs A_k and A_{k+1}. For example, in Fig. 6 P_j falls between P_{j_2} and P_{j_1} but C_j intersects arcs A_1 and A_3. What the location of P_j does imply is that $a \le k$ and $b \ge k+1$. Now, from Lemma 4 we know that at A_a, C_j crosses from being closer to P_i than F to being further, moving counterclockwise. In other words, $e_l \in C_j$ for $l \le a-1$ or $l \ge b$, whereas for $a \le l \le b-1$, $e_l \notin C_j$. Hence the final step of the update procedure consists of a search outwards from k in both directions, based on the test $e_l \in C_j$.

A sequential search will require $O(n)$ in the worst case, since each of the $O(n)$ frontier arcs might be examined before the intersection is detected. However, since every frontier arc examined during the search is deleted forever from the frontier, the search cost for all updates of a wedge as j varies from $i+1$ to n is $O(n)$. On the other hand a binary search will examine at most $O(\log n)$ per search, but might examine $O(n \log n)$ over the lifetime of a wedge. A search procedure that combines the advantages of both is a backward binary search: follow the binary search tree from the rightmost leaf up until a negative test is obtained, and then complete a normal binary search. It is easily verified that this results in an $O(\log n)$ per search worst case with a total search cost of at most $O(n)$.

The algorithm incorporating the sequential search is coarsely specified by the displayed code.

(1) **for** $i=1$ **to** n **do**
  begin {for}
    $j \leftarrow i$
(2)   **repeat** $j \leftarrow j+1$ **until** $P_i \notin C_j$
(3)   Initialize W_{ij}.
(4)   **while** W_{ij} not empty **do**
    begin {while}
      $j \leftarrow j+1$
(5)   **if** $P_j \in W_{ij}$ **then** output $(i,j) \in G$
(6)   Intersect cone determined by $W_{i,j-1}$ and C_j,
        and clip frontier accordingly.
(7)   Determine k such that P_j lies between the centers of
        arcs A_{k+1} and A_k.
      if $e_k \in C_j$ **then** go to (11).
      $a \leftarrow k$
(8)   **while** C_j does not intersect A_a and $a \geq 1$ **do** $a \leftarrow a-1$

      $b \leftarrow k+1$
(9)   **while** C_j does not intersect A_b and $b \leq r$ **do** $b \leftarrow b+1$
(10)   Delete arcs A_a through A_b, and insert A_a', A, and A_b'.
(11)   **end** {while}
  end {for}

ALGORITHM

At the top level, the algorithm consists of two nested loops, the **for** loop on line (1), and the **while** loop on line (4), which together require $O(n^2)$ iterations. Lines (5) and (7) are both accomplished with binary searches, and require $O(\log n)$ time. Lines (6) and (8-10) implement the updating of the wedge frontier, and although any particular iteration may require $O(n)$ time, no more than $O(n)$ work will be required over all $O(n)$ iterations of the **while** loop (4). Thus the average cost for these steps is $O(1)$. The result is a total worst-case time complexity of $O(n^2 \log n)$. Note that only linear space is required.

A final remark concerns the data structure needed to maintain the ordered set of endpoints of the arcs (and their centers) under the above operations. These operations are similar in spirit to the ones needed for the on-line convex hull maintenance, as described by Avis, ElGindy, and Seidel [1]. In particular, the endpoints can be stored in a 2-3 tree structure that is updated using splitting operations. The endpoints are at the leaves of the tree and are at the same time connected among themselves as a doubly-linked list. Every internal node of the tree has, in addition to the tree pointers to its children, pointers to the leftmost and rightmost leaf of the subtree rooted at that node.

Finally we observe that if the algorithm is run "on-the-fly" from $i=1$ to n, although the resulting approximation is not in general the global minimum, the time complexity is only $O(n \log n)$.

Acknowledgments

The second author thanks Heather Booth, Subhash Suri, and Godfried Toussaint for stimulating discussions.

REFERENCES

[1] Avis, D., ElGindy, H., Seidel, R., Simple on-line algorithms for convex polygons, in *Computational Geometry* (ed. G. T. Toussaint) North-Holland, Amsterdam (1985), 23-36.

[2] Imai, H., Iri, I., Computational-geometric methods for polygonal approximations of a curve, *Computer Vision, Graphics, and Image Processing*, **36** (1986) 31-41.

[3] Imai, H., Iri, I., Polygonal approximations of a curve -- formulations and solution algorithms, Kyushu University technical report CSCE-86-C07, 1986.

[4] Toussaint, G. T., On the complexity of approximating polygonal curves in the plane, *Proc. IASTED, Inter. Symp. on Robotics and Automation*, Lugano, Switzerland, 1985.

Computational Morphology
G.T. Toussaint (editor)

UNIQUENESS OF ORTHOGONAL CONNECT-THE-DOTS

Joseph O'ROURKE

Department of Computer Science
Johns Hopkins University
Baltimore, MD 21218 †

It is proven that a collection of non-intersecting simple orthogonal polygons is uniquely determined by its vertex set, and reconstructible from this set in $O(n \log n)$ time. The theorem is also extended to three dimensions: a collection of orthogonal polyhedra is uniquely determined by its set of edges.

1. INTRODUCTION

Our perceptual system's ability to reconstruct the shape of objects from sparse partial data has led pattern recognition researchers to study a particularly pure condensation of the problem: given a set of points (dots) in the plane, connect them with polygonal paths into a "meaningful" whole. This is hopelessly vague as stated, but several natural constraints suggest themselves that lead to interesting problems and solutions [8] [3] [2] [7] [1]. One such constraint requires the dots to be connected to form simple closed polygons. This still leaves so much freedom, however, that it seems difficult to choose a "natural" candidate solution [5]. In search of a more constrained problem, I investigate here the problem of connecting dots into orthogonal polygons.[1] The conclusion is that there is at most one way of connecting a set of dots into a collection of orthogonal polygons. This theorem extends to three dimensions: there is at most one way to flesh out a wire-frame into a collection of orthogonal polyhedra.

More formally, let P be a collection of non-intersecting simple orthogonal polygons in the plane. Each polygon is composed of an alternating sequence of horizontal and vertical edges, any two of which meet in at most one point, which point is a vertex of the polygon. Each edge contains

† This research was partially supported by grants from NSF (DSR 83-51468) General Motors, IBM, and Martin Marietta. An earlier version of this paper was released as [4].
[1]First suggested by Derick Wood.

exactly two vertices; vertices are only permitted at corners. The polygons may be nested inside one another, but they may not intersect or share vertices. The vertices of the polygons are not necessarily in "general position": any number may fall on the same vertical or horizontal line. We will view P as composed of two sets: the vertices V and the edges E.

Now imagine erasing all the edges of P, leaving only the set of vertices V on the plane. The main result of this note is that E is uniquely determined by V. The curious implication is that the edges of orthogonal polygons are superfluous: there is only one way (if there is a way) to connect a set of dots into orthogonal polygons. Moreover, the proof of this theorem is constructive, leading to an $O(n \log n)$ algorithm for reconstructing P from V.

2. TWO DIMENSIONS

Each vertex is incident to exactly one horizontal edge and one vertical edge. A vertex that is leftmost on the horizontal line containing it must be the left endpoint of a horizontal edge, and a vertex that is uppermost on the vertical line containing it must be the upper endpoint of a vertical edge. These simple observations lead to the following algorithm.

Index the vertices on a particular horizontal line H from 1 to m; m must be even because each horizontal edge on H has two distinct endpoints. Let $[i,j]$ be the horizontal edge from vertex i to j on H. Then the edges $[2i-1,2i]$ for $1 \leq i \leq m/2$ must be in E, and these are the only edges in E that lie on H. $[1,2] \in E$ because 1 is leftmost; $[2,3] \notin E$ because 2 may only have one incident horizontal edge; $[3,4] \in E$ because 3 must have an incident horizontal edge; and so on. Therefore all the horizontal edges of P may be reconstructed by connecting alternate pairs of dots in each horizontal row as above. Similarly all vertical edges can be reconstructed by connecting alternate pairs of dots in each vertical column from top to bottom. Fig. 1 shows a particular point set and Fig. 2 shows its unique reconstruction.

This procedure can be accomplished easily in $O(n \log n)$ time by sorting the vertices horizontally and vertically.

Note that the algorithm also detects when a point set cannot be the vertex set of a collection of non-intersecting simple orthogonal polygons: whenever no move can be made at a step of the algorithm, or when edges determined by the algorithm cross. An example of the former case is shown if Fig. 3a, and of the latter case in Fig. 3b.

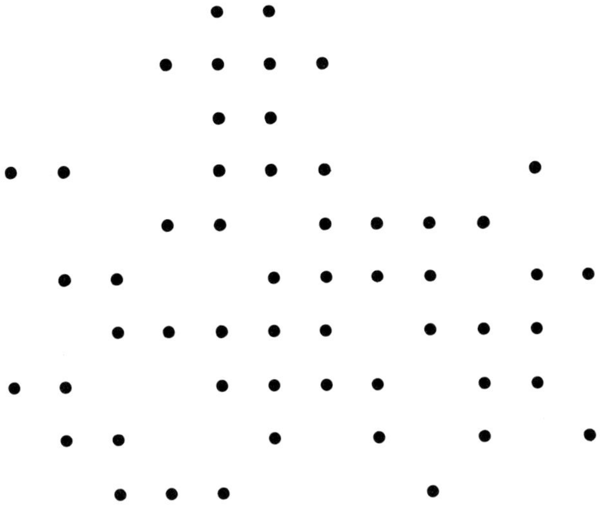

FIGURE 1
A set of dots in the plane.

FIGURE 2
The orthogonal polygons reconstructed from Fig. 1.

The results of preceding discussion can be summarized in the following theorem:

THEOREM 1. A collection of non-intersecting orthogonal polygons is uniquely determined by its set of vertices V, and its edges can be reconstructed from V in $O(n \log n)$ time, where $|V| = n$.

3. THREE DIMENSIONS

This theorem cannot be extended directly to three (or higher) dimensions: two identical crosses arranged parallel, one above the other, can be connected by a central beam without addition of vertices. More specifically, reflect the three points { (2,1), (1,1), (1,2) } in the X-axis, the Y-axis, and the origin, resulting in 12 points defining a cross in the XY plane. Define V to be the 48 points composed of this pattern at $Z=0$, $Z=1$, $Z=2$, and $Z=3$. Then the central rectangular box defined by the square { (1,1), (-1,1), (-1,-1), (1,-1) } between $Z=1$ and $Z=2$ either may or may not be present; see Fig. 4. Thus V does not uniquely determine a collection of orthogonal polyhedra.

The crucial property used in the two-dimensional proof that no longer holds in three dimensions is that each vertex is incident to exactly one horizontal and one vertical edge. Call an edge parallel to the X-axis an X-edge, and similarly for Y- and Z-edges. Then in the $X=1$ plane, the vertex (1,1,1) in Fig. 4 has an incident Y-edge [(1,1,1), (1,2,1)], but may or may not have the incident Z-edge [(1,1,1), (1,1,2)], shown dotted in Fig. 4.

To resolve this ambiguity, it is necessary to know the edges as well as the vertices. Of course the edges determine the vertices by definition, so only the edges need be given for reconstruction. We now establish this claim.

Let P be a collection of simple orthogonal polyhedra in three dimensions. Each polyhedron is composed of faces that lie in planes parallel to two of three orthogonal coordinate axes X, Y, and Z. A plane parallel to the Y and Z axes is called a YZ-plane, and a face in the YZ-plane is called a YZ-face, and similarly for the other two combinations. Each face is an orthogonal polygon. Each edge on the boundary of a face is contained in exactly two orthogonal faces; note that there must be a right angle at each edge: edges are not permitted to lie at the junction of two coplanar faces. Distinct faces intersect only at edges and vertices, if at all. We view P as composed of three sets: the faces F, the edges E, and the vertices V.

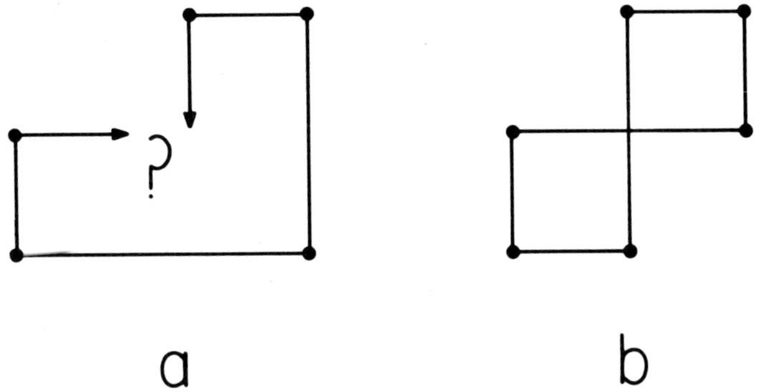

FIGURE 3
Two illegal reconstructions: (a) no matching vertex; (b) illegal self-intersection.

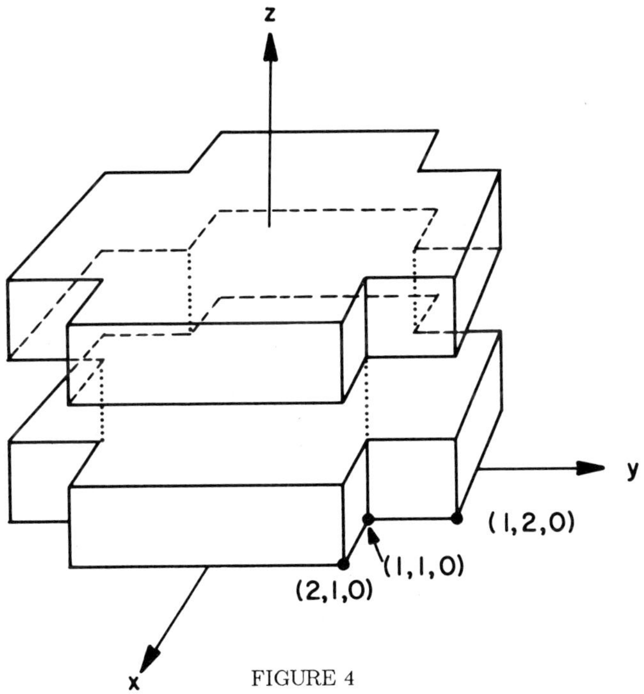

FIGURE 4
An example of 48 vertices in three dimensions that do not uniquely determine polyhedra: the central box between the two crosses (whose vertical edges are dotted) either may or may not be present.

Suppose we are given E, with each edge defined by a pair of endpoints. V is simply the union of all these endpoints. We reconstruct F from E in three parts. The first step computes the YZ-faces. Let $X=c$ be any YZ-plane π that wholly contains at least one edge $e \in E$. Since e is in the boundary of exactly two orthogonal faces, one of them, say $f \in F$, must lie wholly in π. Recall that f is an orthogonal polygon; so all its boundary edges lie wholly in π.

It is natural then to conclude that we have an instance of the two-dimensional problem just solved: a collection of orthogonal polygons in π. But the collection does not necessarily satisfy the assumption that each vertex is incident to exactly two edges. Returning again to Fig. 4, let π be the plane $X=1$. Then, if the central box is absent, the collection of edges in π is as shown in Fig. 5a; if the box is present, then as in Fig. 5b. In Fig. 5b, some vertices are incident to four edges.

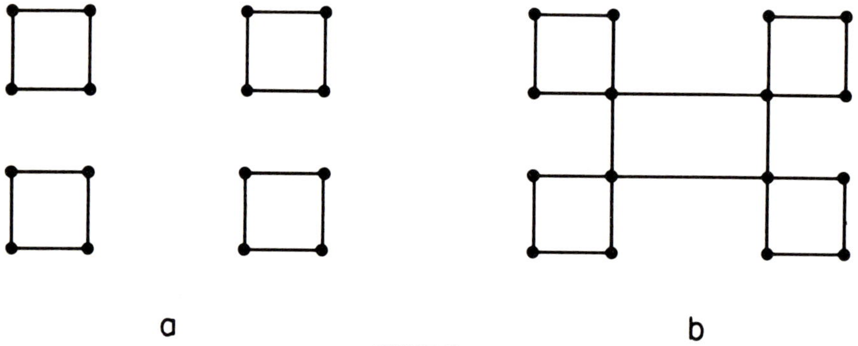

FIGURE 5
The edges that lie in $X=1$ in Fig. 4 if the central box is not (a) or is (b) present.

Nevertheless it is not difficult to reconstruct the faces in π from the edges in π. Let H be a horizontal line in π parallel to the Y-axis that intersects at least one edge but contains no vertices. Index the edges intersected by H from 1 to m. Now we have a variant of the problem solved previously. Let $[i,j]$ represent the closed segment on H from edge i to edge j. Again m must be even, and the segments $[2i-1,2i]$ for $1 \leq i \leq m/2$ must be subsets of faces. Sorting the vertices in π vertically by Z-coordinate and locating H between every adjacent pair in this sort, is clearly sufficient to identify the interior of each orthogonal face. See Fig. 6.

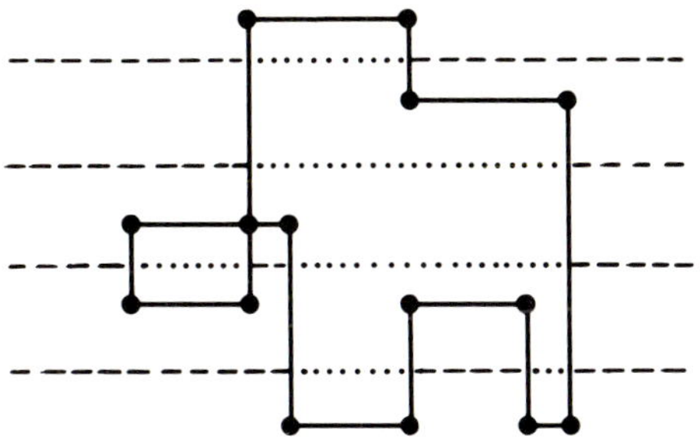

FIGURE 6

The interior of the faces may be detected by positioning a line between vertices; here dashed indicates exterior and dotted interior.

Reconstruction of all YZ-faces can then be achieved by setting π to each YZ-plane that contains an edge. The XZ- and XY-faces may be reconstructed similarly. Informally, the conclusion of this argument is that orthogonal polyhedra may be reconstructed from their "wire-frames." The following theorem is a more formal statement.

THEOREM 2. A collection of orthogonal polyhedra that intersect only at vertices (if at all) is uniquely determined by its set of edges E, and its faces may be reconstructed from E in $O(n \log n)$ time, where $|E| = n$.

Note that the polyhedra may intersect at vertices, as this does not destroy the crucial property that each edge is shared by exactly two orthogonal faces.

4. DISCUSSION

The generalization of this theorem to higher dimensions is that a collection of orthogonal polytopes in d-dimensions is uniquely reconstructible from its set of $(d-2)$-dimensional facets or "ridges." The proof of this claim is similar in spirit to those above and will not be detailed. Note that in higher dimensions, the number of these facets is no longer necessarily linear in the number of vertices. Perhaps of more interest is a question posed by Raimund Seidel: can the two-dimensional theorem be generalized to polygons with edges parallel to a given fixed set of directions? Exploration of this question may uncover a middle ground between the extreme freedom of the unconstrained problem, and the total lack of freedom of the orthogonally constrained problem.

In another direction, David Rappaport recently showed that if the assumption that each vertex is a corner is removed from the two-dimensional problem, thereby permitting two collinear edges to meet at a vertex, the reconstruction problem becomes NP-complete [6].

Acknowledgements

I thank Michael McKenna for insightful comments. This research was partially supported by NSF Grant DSR83-51468, and grants from General Motors, Martin Marietta, and IBM.

REFERENCES

[1] Ahuja, N., Dot pattern processing using Voronoi neighborhoods, *IEEE Trans. on PAMI,* **PAMI-4** (1982) 336-343.

[2] Fairfield, J. Contoured shape generation: forms that people see in dot patterns, *Proc. IEEE Conf. Sys., Man, Cyb.*, (1979) 60-64.

[3] O'Callaghan, J. F., Computing the perceptual boundaries of dot patterns, *Comput. Graphics and Image Proc.*, **3** (1974) 141-162.

[4] O'Rourke, J., Reconstruction of orthogonal polygons from vertices, Johns Hopkins tech. report JHU/EECS-85/13 (1985).

[5] O'Rourke, J., Booth, H., Washington, R., Connect-the-dots: a new heuristic, *Computer Vision, Graphics, and Image Processing*, **39** (1987) 258-266.

[6] Rappaport, D., Computing simple circuits from a set of line segments is NP-complete, *Proc. Third ACM Symp. Computational Geometry*, Waterloo, (1987) 322-330.

[7] Toussaint, G. T., Pattern recognition and geometric complexity, *Proc. 5th Int. Conf. Pat. Recog.*, Miami Beach (1980) 1324-1347.

[8] Zahn, C. T., Graph-theoretical methods for detecting and describing gestalt clusters, *IEEE Trans. Comput.*, **C-20** (1971) 68-86.

Computational Morphology
G.T. Toussaint (editor)

ON THE SHAPE OF A SET OF POINTS

John D. Radke*

Inspired by recent developments in computational morphology, this paper discusses their potential impact on research undertaken in the field of spatial analysis addressing the characterization and recognition of form in point sets. A brief discussion of point pattern recognition methods now common in spatial analysis is included, pointing to their limitations and questioning their success. The main focus of the paper, however, is the examination of recent methods of geometric decomposition that appear more useful for solving questions concerning form. New techniques based in computational morphology may very well revolutionize the characterization of point sets for spatial analysts. Some of these techniques are referenced and briefly discussed here, including their potential applications to point pattern recognition problems in spatial analysis.

1. Introduction

What at first appears to be a simple task, to address the shape of a set of points, becomes a complex and thought provoking endeavour when the points are locational identifiers of some real spatial phenomenon. Physical, biological and social scientists have always been faced with this challenge but progress seems to have been made in the past few decades. This paper identifies the use of some ideas and techniques of computational morphology that appear useful to scientists in a variety of disciplines.

Computational geometry is quickly expanding and influencing other fields of study. Computational morphology, according to Toussaint occurs when "a computational geometric structure is intended to extract the form of the input" [T3]. The input, usually representing a spatial distribution of some phenomenon, can be easily characterized by a set of elements, such as points, lines or areas. The recognition of spatial patterns within these

----

* Department of Landscape Architecture and Regional Planning, University of Pennsylvania, Philadelphia, PA 19104-6311.

element sets often leads to further understanding of generative processes involved in their construction. If we consider the recognition potential of extracting and characterizing form, it is clear that computational morphology is a field in itself.

Spatial analysis has been identified by scientists as a field of study which questions and examines the spatial structure and association of phenomena. Computational morphology, although a relatively new concept, has become increasingly important to the ideas and tools employed in spatial analysis. Its greatest impact has probably been the unification of ideas concerning geometrical properties of structures that are fundamental to a variety of pattern recognition problems. Work on geometrical properties, some of which has already had an impact on spatial analysis, is addressed by Toussaint [T3].

The central objective of this paper is to examine decompositions of point sets that can better characterize the morphological structure of "real world" point patterns. It is argued that the traditional methodologies of spatial analysis are limiting and are often insensitive as descriptors of form. This paper is inspired by several works in computational morphology which employ conceptually similar techniques and ideas concerning the spatial association and characterization of point patterns (cf. [EKS,KR]).

To characterize completely some phenomenon in space would in general involve a variety of analytical techniques. The purpose of this paper is not to dwell on the holistic approach, but rather to identify contributions made to the recognition process in the field of spatial analysis by computational morphology. Therefore, attention is focused on geometric properties of point sets which are embedded in a plane.*

In order to understand the role computational morphology has played and will continue to play in the field of spatial analysis, we must understand of the reason that we extract and characterize form. The next section of this paper explores the quest for form in spatial analysis. Section 3 details the

----

* The study of point patterns embedded in a plane is most common in spatial analysis (cf.[GB,HCF]).

properties of point sets often examined in spatial analysis; some methods for this are described in Section 4. This is followed by some notions of the shape of point sets while Section 6 discusses techniques of computational morphology which look promising for spatial analysis. The paper concludes with a discussion on further influences that computational morphology may have on spatial analysis.

2. Motivation for Studying Form

To explain spatial variance of phenomena is one of the reasons that both physical and social science studies it. The [illegible]ure of the physical sciences has meant they have been more successful in explaining spatial variation than have the social sciences who deal with living systems. Although the two sciences are fundamentally different, common philosophical approaches to explanation exist and have flourished for several decades.

The key to explanation lies in the understanding of the process-form relationships that drive the generation and distribution of phenomena under study. If the underlying process that gives way to the observed form is determined, than changes in form can be explained and prediction is possible. However, in real world situations, the underlying process that is responsible for the distribution of some phenomenon is often not known and only information regarding its form can be obtained. However, description of a distribution's form can prove valuable in the quest for explanation.

Data describing a distribution's form fall into two categories: one which accounts for its aspatial characteristics, such as attribute information (i.e. describing population of a city), and the other which summarizes its spatial characteristics. Of interest to us are the spatial characteristics that can be extracted from a distribution as they relate to computational morphology.

Of the two benefits arising from the study of form, one is descriptive, while the other is more analytical. The description of form can vary widely, from simple qualitative notions to more vigorous and quantitative approaches. In the field of spatial analysis, to summarize complex patterns has become one of the main reasons for the descriptive approach (cf.[BG]). Here a high degree of description is sought so that spatial structure can be characterized. In exploratory pattern analysis, spatial structure may at first go undetected but it may be recognized with further and more rigorous study (cf.[SJ]).

The second benefit and reason for studying form is associated with identifying the processes that give rise to the form being examined. With more knowledge about the spatial characteristics underlying the form of a point set, we can hypothesize and focus more clearly on what the generative process may be. In such circumstances, the recognition of form is both an end in itself, and a means to explain or predict by linking process and form (cf.[BG]).

3. Properties of a Point Set

Before discussing methods used in point pattern analysis, it is necessary to review some basic properties of point sets. For most scientists studying real world problems, gathering data is a complicated prospect. In order to address the problems effectively, researchers filter through vast amounts of information and remove 'noise' by extracting only data considered to be essential to the problem. Studies in spatial analysis have popularized this approach and essential data has been stored in a form which emphasizes its spatial features. This form is commonly known as a map and is a two dimensional representation of a three dimensional world.

Taking the abstraction one step further, researchers have used graph theoretic concepts to build a framework in which to study problems of morphology. In this framework objects in the real world are stored as symbols which represent their geometric characteristics. For example, phenomena being examined could be represented by points, lines or areas.

Although a point is considered dimensionless, real world objects are represented as points when the distances between them are considered important and their physical size is irrelevant. Similarly, information is stored as a line when the relationship between two points defining the line and not the characteristics of the line itself is essential. Often a set of points can be further divided into subsets, and members belonging to different subsets are dissimilar. The result of this process, commonly referred to as clustering (cf.[JBD,GK]), defines an areal feature commonly known as a polygon.

Haggett [H] introduced a paradigm (popular in spatial analysis) which summarizes relationships between graph theoretic structures such as points, lines and areas. The relationships are based on an element's dimensionality and the point, line and cell (the term used by Haggett to represent area) are separated into different classes. Much of the research in spatial analysis has focused on individual classes with little attempt made to unify or generalize solutions. Computational morphology, however, appears to be able to generalize form recognition solutions (cf.[KR]). Bringing spatial analysis and computational morphology together is clearly one of the motivations for this paper.

Since all objects in their most elementary form can be represented spatially as sets of points, studying point sets is the logical first step in characterizing form. Much of the empirical research has extracted and abstracted elements from reality which can be represented as points embedded in a plane. For simplicity, the methodologies examined in this paper will address the recognition of the forms of point sets in the plane. Many of these methodologies have analogs in higher dimensions; however, to address these, however, would go beyond the intent of this paper. For the remainder of this paper, let S denote a fixed set of n points in the plane.

4. Methods of Point Pattern Analysis

An indepth look at methodologies that have been used to examine point patterns would prove distracting here and would not serve this paper's end. However, it is important to mention the approaches now common to point pattern analysis. These approaches will not be examined in detail; rather, their limitations at characterizing spatial form will be emphasized. Advances in computational morphology, which are detailed in Section 6, look able to overcome many of the deficiencies pointed to in this section.

Several classifying techniques have been used to study point patterns (cf.[HCF,CO,GB]]. The majority of techniques involve quantitatively summarizing the overall structure of a point set rather than identifying specific characteristics of the elements that comprise it. There are two basic categories of techniques, those dealing with absolute measurements of the dispersion of points about the area occupied by a point set, and those examining the association of points relative to other points in the set. The former category includes measures of density, distance and direction, while the latter includes relative measures associated with point neighbours. Although the term "neighbour" is used here, the emphasis is on point-to-point relationships; areal ideas of neighbourliness, common in computational morphology, are not considered.

The approaches all employ a typology based on logical procedures in testing the validity of a hypothesis. In order to identify the form under study, it is useful to compare it to a benchmark whose properties are well known. In this instance, the characteristics of the observed point set are statistically compared to those expected, given that the observed form is a result of a Poisson process (cf.[HCF]). Measurements are recorded as variations from the benchmark and are separated into three basic categories: random (reflecting properties of a Poisson process), clustered, and dispersed or regular [Ta].

Quadrat methods, popular among plant ecologists [Ta], calculate dispersion by summarizing a point set in terms of the density of points per quadrat. A quadrat is defined here as a basic areal unit into which the study region is divided. The observed frequency distribution is compared to the expected bench mark which is generated by the theoretical Poisson and a statistical goodness-of-fit test is carried out. A test of variance-mean ratio reveals whether the form of a point set is dispersed, clustered or like one which would be generated by a Poisson process.

The quadrat methods have limitations which cast doubt on their results. The results are influenced by the number and size of the quadrats used and as they are altered, the density of the distribution is altered. This can radically change the observed frequency distribution and affect the Poisson probabilities and thus the results. The method is considered a global metric as it summarizes the whole pattern. Frequency distributions of two point sets that vary internally may thus not appear to differ when examined using this method. However, when visually examined they appear obviously different (Figure 1).

Finally, the quadrat methods are plagued with boundary problems that have not yet been solved (cf.[Ta]). The properties of the theoretical Poisson benchmark assume a fixed density of points distributed over an infinite plane. However, observed real world point sets are bounded and the process used to delineate their boundaries often mean density is miscalculated in the method. As well, an incorrect boundary can eliminate points that belong to the point population under study or include points that belong to another population. The boundary problem itself is not a trivial one and many attempts have been made to correct for it (cf.[GA,Ri]). Where appropriate, problems associated with the bounding of a point set will be discussed in this paper.

A method calculating dispersion and based on a distance measurement summarizes point sets using a nearest neighbour statistic. This method, also based upon the Poisson distribution, was first developed by astronomers in the late nineteenth century and made popular by ecologists in the 1950s [CE]. The frequency distribution calculated refers to distances to nearest neighbours, rather than the observed frequency distribution representing points per area, as in the quadrat method. Mean nearest neighbour distances are calculated for both the observed point set and expected Poisson distribution. An index generated from the observed - expected mean ratio summarizes the pattern as either dispersed, clustered or similar to a Poisson pattern.

Although the nearest neighbour method is not plagued with quadrat size problems, it too can be altered by density changes as density is critical to the Poisson frequency distribution. The delineation of a boundary determines the density of the observed point set, but he Poisson frequency distribution assumes an unbounded point set. This method is also influenced by the boundary edge itself while calculating nearest neighbour distance. Some boundary edges may be closer to the point under

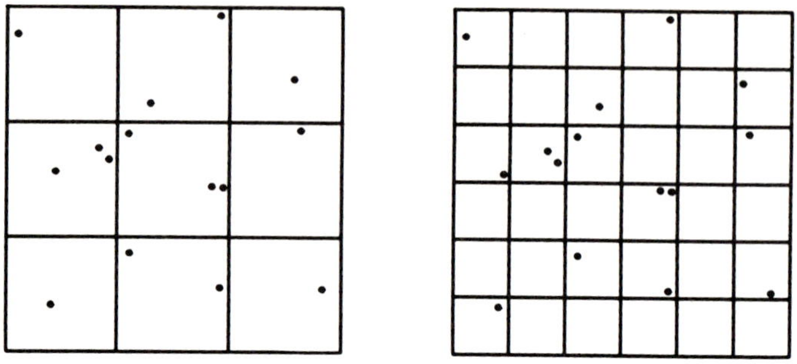

# of points	frequencies
0	-
1	5
2	2
3	2

# of points	frequencies
0	23
1	11
2	2
3	-

Figure 1 QUADRAT SIZE EFFECTS ON FREQUENCY DISTRIBUTION

observation than its observed nearest neighbour. A number of attempts to solve this problem (cf.Ta,GB]) have resulted in distance bias. There are other boundary problems that exist here that are similar to those discussed for the quadrat method above. In addition, reflexive neighbours (where two points are each other's neighbours) cause the same distance to be measured twice resulting in a biased estimate.

There are several descriptive summary measures using distance and central tendency which have their roots in centrography. This method parallels descriptive statistics common in aspatial data and includes area measures for the median and mean. The standard distance deviation can be calculated and plotted to illustrate spatial dispersion in the point set. Although this method does not have the density limitations of the approaches above, it does have boundary problems similar to those already discussed.

A dispersion metric based on distance and with directional tendencies was first introduced to summarize sociological point sets by Lefever in 1926 [Le]. This statistic, the Standard Deviational Ellipse, calculates the areal mean of the point set and then proceeds to maximize the standard distance deviation along one axis while minimizing it along a perpendicular axis. If an exact solution exists (cf.[R1]), the major and minor axes are generated, and the standard deviation bounding ellipse can be plotted. This method, like the standard distance deviation, has no density constraint and is affected by the boundary imposed on the observed point set. Although it directionally orients the dispersion about the mean, its global summarization has been criticized as being too insensitive for recognition purposes.

An expanding body of research employing methods measuring the association of points relative to each other is proving to be less problematic than the dispersion methods just discussed. These methods are free from problems associated with pattern density and problems generated by edge effects. However, these methods still attempt to summarize the whole point set and although some effort is made to measure within a neighbourhood, localized sub-patterns within the point set are not sought.

Reflexive or reciprocal nearest neighbours can be observed and compared to those expected in a Poisson generated pattern (cf.[CE]). Although there is no statistical test available to evaluate significance here, the method is useful in summarizing a point set if higher order neighbours are included. This involves generating frequency distributions representing the number of second, third, fourth,... reflexive nearest neighbour counts and comparing them to those expected in a theoretical Poisson. As in the quadrat and nearest neighbour methods evaluating dispersion, the point set is categorized into three types: regular, clustered, or that similar to a Poisson generated pattern. This method can be applied to a point set with linear rather than areal properties (cf[Da]), but it can be misleading when the number of points in the set is low [PW].

The final method examined uses both a polygonization and triangulation method which appears more promising than either the reflexive neighbour or dispersion methods. The polygons generated here are properly called Theissen or Voronoi polygons. The Delaunay triangulation (DT), which is critical to this method, is the dual of Voronoi polygons, which implies that the edges of the DT join Voronoi neighbours [SH]. Two points x and y are Voronoi neighbours if there exists a point z in the plane for which

$$d(x,z)=d(y,z)=\min\{w,z)\,|\,w \in S\}.$$

Toussaint [T3] provides a good review of algorithms that can generate the Voronoi diagram. Shamos and Hoey [SH] describe an O (n log n) algorithm which employs a divide and conquer approach for constructing the planar Euclidean Voronoi diagram. Once the Voronoi diagram is computed, the DT can be computed in O(n) time and thus O(n log n) in total [T3].

Although the DT has many well-explored properties, the sizes of the angles between DT edges have been used by spatial analysts to characterize form. Specifically, the minimum angle of each triangle is measured and added to an observed frequency distribution. This can then be compared to a expected frequency distribution to summarize the point set [MEP].

The scale independence of this method is its strong point. However, measurements are still global summaries of the point set's spatial structure and other valuable properties of the DT have not been exploited. Many questions regarding the form of point sets still remain unanswered.

Both the Voronoi polygons and the Delaunay triangulation had unifying effects in computational geometry. It is interesting that they appear in this last attempt to characterize form. Unfortunately, comparing an observed frequency distribution to an expected theoretical Poisson in order to generate a gross global summarization does not capitalize on their true strengths. More appropriate approaches to form recognition that have roots in computational morphology will be shown.

5. Notions of Shape

The shape of point sets has not been as widely discussed as dispersion or association within point sets. In fact, what is meant by shape lacks an explicit answer [KR]. Toussaint [T3] points out that human observers can perceive shape in planar point sets which have a finite diameter (so they are visible) and are fairly densely distributed. He describes these point sets as dot patterns or dot figures with a bounding shape decomposition, the shape hull.

The decomposition of a point set into components that identify "essential" points tends to generate a structure that can be considered a boundary (similar to shape hulls discussed by Toussaint [T3]). Although the term "essential" can refer to several qualities of a point set, in the study of shape, the essential points are considered those with qualities (due to their spatial association within the point set) that can be used to decompose the point set to a perceptually meaningful form detecting its geometrical properties.

The decomposition of a point set can result in either a linear or polygonal (area) characterization of its shape. These structures identify aspects of shape that for the sake of clarity can be classified as internal or external shape (cf.[KR]). Toussaint [T3]) uses the term "skeleton" to describe the decomposition to an internal line structure of a set of points, while he defines the external polygonal decomposition as the shape hull. Pavel [Pa] uses the internal or external skeleton as being a "generalized symmetry axia of the image, obtained by taking away points of the image while preserving all the useful information".

Whether the shape of the point set is characterized by an internal skeleton or an external hull, the idea of association between points is important. This association can be considered being neighbourly (cf.[T3,KR]). In the case of the shape hull the association is between "essentially" extreme points that lie on the periphery of the point set (they can be considered global neighbours) and in the case of the internal "skeleton", association is between "essential" points that are near in proximity (they can be considered local neighbours).

The descriptors of form to spatial analysis are found to produce solutions which, for the most part, are global summarizations of spatial structure. These summarizations are too insensitive to be considered sufficient descriptors of form and a local association and summarization of points is sought. In addition, Haggett's [HCF] paradigm has served as a conceptual barrier keeping separate the recognition solutions for point, line and area patterns. The shape metrics discussed in this next section look promising and appear to overcome many form recognition problems now plaguing spatial analysis.

6. Decompositions That Characterize Form

Although Toussaint [T3] points out that the amount of progress made in "geometrical pattern recognition" has been staggering, there is much work to be done. Decompositions of point sets or some parameterized notion defining sets, such as a skeleton or a polygon, into simpler components result in better morphological descriptions. These descriptions of form or "anthromorphic" decompositions [P] appear more perceptually meaningful than the analytical techniques discussed that are common in spatial analysis.

There is no one descriptor of form that is best for all occasions. The problems we attempt to solve vary in complexity and so do the point patterns involved in their solution. As a result, the available descriptors of form also vary in complexity. Following Kirkpatrick and Radke [KR], an examination of shape descriptors is divided into external and internal measures. Within these two categories, simple decompositions of point sets that summarize either global or local characteristics are examined and their ability to accomplish both global and local summarization.

External Shape Descriptors

The minimum bounding box, the minimum bounding circle [FS,TB] and the minimum bounding ellipse (a generalization of the minimum bounding circle [KR]) are simple decompositions describing external shapes in point sets. The next step in descriptive complexity is the convex hull which, except for degenerative cases, yields a fairly complex polygonal description of the external boundary of a point set [AT,T1,T3]. The convex hull, by definition, summarizes only the global shape of a point set. Although the convex hull may include unnecessary points and result in too detailed descriptions of form [KR], it is more likely that the decomposition will not include "essential" points that exist on the periphery of the point set.

Edelsbrunner, et al, [EKS], overcome the global restrictions of the convex hull by introducing the notion of a shape spectrum which seems to capture the intuitive notion of "finer" or "cruder shape". This family of hulls, called here α-hulls, is a generalization of the convex hull.

Given a set S of n points in the plane, its α-hull is the intersection of all closed discs with radius $1/\alpha$ that contain all n of S. For arbitrary real valued α, a generalized disc of radius $1/\alpha$ is defined. When $\alpha=0$, the disc is a half plane which generates the convex hull of S. If $\alpha>0$, the disc has a radius $1/\alpha$ and generates a global description of shape which tends to be rather crude when α is "relatively large" [EKS]. Since at least one disc must exist that contains all n in S, the smallest possible value of $1/\alpha$ is equal to the radius of the minimum bounding circle. For $\alpha < 0$, the complement of a disc with radius $1/\alpha$ is used to construct the hull which is called a negative α-hull. The family of α hulls exists for α ranging from $-\infty$ to ∞ . The smaller the value of α, the more localized and sensitive the descriptor of shape.

Edelsbrunner, et al, also define a combinational variant of the α-hull, called the α-shape of a point set. The α-shape represents the relationship between the "essential" points in the point set as they are detected by the enclosing α-hulls. Unlike the family of α-hulls, the family of α-shapes has a finite number of members. A member of the α-shape family can be generated and remain stable (even though the value of α changes), until the set of "essential" points detected by the α-hulls changes.

It is clear that in a very complex point set S, the generation of α shapes will aid in the decomposition of S into meaningful polygonal subsets which can characterize spatial relationships between points in S at a "global" as well as "local" level. This generalized notion of α better characterizes the external shape of a point set than does the discrete hull constructs mentioned earlier. The notion of shape spectrum introduced by Edelsbrunner, et al, appears much stronger than the element counts and subsequent frequency distribution tests undertaken in spatial analysis.

Internal Shape Descriptors

A discussion of internal descriptors of form can easily parallel that addressing external descriptors. Like their external counterparts, simple decompositions of point sets that summarize global or local characteristics have enjoyed widespread use as descriptors of shape (cf.[T3]). Further decomposition of the same point set can yield a variety of descriptors increasing in complexity and their ability to accomplish both local and global summarization.

There exists a discrete family of nested and progressively more detailed decompositions that have gained much attention as internal descriptors of form [KR]. The process underlying their construction uses the notion of neighbourliness to identify and connect "essential" points in the set (those points that are fundamental in the link generation process.) These internal descriptors can be thought of as forming the endo-skeleton shape of a point set. In contrast, α-shapes can be thought of as constructing the exo-skeleton of the point set.

The most rudimentary realistic descriptor of internal structure in the discrete family of decompositions is the set of all nearest neighbour pairs. The skeleton generated when all nearest neighbours are linked is the nearest neighbour graph (NNG). The next member of the family, the (Euclidean) minimum spanning tree (MST), is a supergraph of the NNG. The MST includes all the nearest neighbour parts while connecting all members of the point set and at the same time minimizing the resultant graph's total edge length. Shamos, et al, [SH] present an O(n log n) algorithm for both the NNG and the MST which are known to be asymptotically optimal in the worst case.

Although the MST has had widespread use as a descriptor of a point set (cf.[Z]), the next member of the family, the relative neighbourhood graph (RNG) "appears to be superior" [T3]. Toussaint [T3] describes the generation of the RNG by defining the criteria for which points in the set are viewed as "essential" and thus relative neighbours. Two points x and yare relative neighbours in S if

$$d(x,y) \leq \min\{\max\{d(x,z),d(y,z)\}|z \in S\}.$$

Toussaint discusses several algorithmic issues (cf.[T2,TM]) and an O(n log n) algorithm exists for the RNG [Su].

The next graph in the hierarchy of discrete decompositions is conceptually similar to the RNG and is commonly known as the Gabriel graph (GG) [MS]. The graph can best be described by examining this relationship between its "essential" points. Two points x and y are Gabriel neighbours in S if

$$d(x,y) \leq \min\{\{d^2(x,z) + d^2(y,z)\}^{\frac{1}{2}}|z \in S\}.$$

Matual and Sokal [MS] suggest an O(n log n) algorithm.

It is trivial to see that all the preceding discrete graphs are subgroups of the Delaunay triangulation (DT) defined earlier in Section 4. This is important from a descriptive point of view, as well for construction. As a descriptor, the DT is the maximal planar description if internal structure. When construction issues are discussed, all subgraphs can take advantage of the DT being generated in O(n log n) time.

Figure 2 illustrates the discrete family of graphs for a fixed point set. As descriptors of shape, the graphs discussed here are all scale invariant. The discrete family of graphs can also be used to analyze local spatial properties and neighborhood relationships between individual elements in a point set. These characteristics alone provide a more universal application of the graphs as point set descriptors.

Following the use of the DT in spatial analysis (cf.[GB,MEP]), there is immediate potential with other members of the discrete family to extract and statistically analyze global properties. As previously mentioned, this method can be criticized for its simplicity as the decision of similarity is based on a comparison between measured characteristics of an observed structure and one of the members of the discrete family. This simplicity parallels the comparison to a regular, clustered or Poisson bench mark in spatial analysis. What is lacking is a sensitive method of comparison which can relate each individual element of the observed point set to a theoretical set with known parameters.

Kirkpatrick and Radke [KR] have developed measures of internal shape (neighbourliness) that form a continuous spectrum. The hierarchy of internal shapes that serve to decompose a point set are called β - skeletons and are based on a parameter of neighbourhood. The set edges joining all β - neighbours in S defines the β - skeleton of S. The technique for generating the skeletons is conceptually similar to and was inspired by work describing the external shape of a point set, α - shapes [EKS]. Urquhart [U] also uses a similar approach in developing clustering algorithms.

Kirkpatrick and Radke [KR] define a continuous family of neighbourhoods indexed by a single positive real valued parameter β. Urquhart [U] develops a region of influence with similar characteristics. Simply, two points x and y are β neighbours in a point set S if their neighbourhood or region of influence (defined as $N(x,y,\beta)$) contains no point of S other than x and y. While distance has often been used to define neighbourliness, with scale changes and density variations, attempting to compare structure within or between point sets (as pointed out in Section 4) can be problematic. Distance methods thus remain insensitive

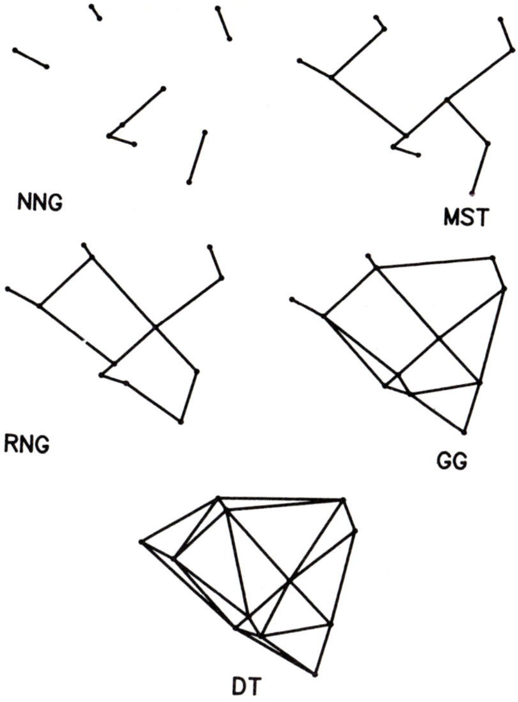

Figure 2 (FROM [KR]) NNG, MST, RNG, GG AND DT OF A FIXED PRINT SET

to important aspects of internal structure such as intervening opportunities created by relative locations of nearby points.

Keeping these intuitive notions of "friction" or "influence" of nearby points in mind, Kirkpatrick and Radke [KR] de-emphasize the role of distance for more scale-independent decompositions that take advantage of the geometrical properties of point sets. Although it is likely that numerous alternatives exist, Kirkpatrick and Radke [KR] present two methods of generating indexed neighbourhoods based on existing and well studied notions of neighbourhood. The first attempts to subsume the discrete family of graphs discussed earlier by using a "lune" (conceptually similar to the RNG's "lune" [T3]) to delineate the neighbourhood. For $\beta > 1$, $N(x,y,\beta)$ is the intersection of two circles of radius $\beta(x,y)/2$ centered at points $(1-\beta/2)x+(\beta/2)y$ and $(\beta/2)x+(1-\beta/2)y$. When $\beta=1$, $N(x,y,\beta)$ corresponds to the Gabriel graph "circle of influence" and when $\beta=2$, the relative neighbourhood graph "lune" is generated.

When β $[0,1]$, $N(x,y,\beta)$ is the intersection of two circles of radius $d(x,y)/(2\beta)$ passing through both x and y. Again, when $\beta=1$, $N(x,y,\beta)$ corresponds to the Gabriel Graph "circle of influence". Figure 3a illustrates the neighbourhoods $N(x,y,\beta)$ for various values of $\beta>0$.

The second method of generating an indexed neighbourhood is circle based and similar to the notion of empty discs used to generate the formalized shape spectrum introduced in α - shapes [EKS]. When $\beta[0,1]$, the definition of the neighbourhood is equivalent to the "lune" based method, $N(x,y,\beta)$ is the intersection of two circles of radius $d(x,y)/(2\beta)$ passing through both x and y. Again, when $\beta=1$, $N(x,y,\beta)$ corresponds to the Gabriel Graph "circle of influence". However, when $\beta>1$, $N(x,y,\beta)$ is the union rather than the intersection of the two circles of radius $\beta d(x,y)/2$ that pass through x and y. As β approaches ∞, $N(x,y,\beta)$ approximates the entire plane and resultant skeleton is null. Figure 3b illustrates some members of the family of circle-based neighbourhoods. Kirkpatrick and Radke [KR] discuss the complexity of constructing the complete β - skeleton and conclude that it can be constructed in $O(n^3)$ time and when $\beta>1$, the worst case is $O(n^2)$ [KR].

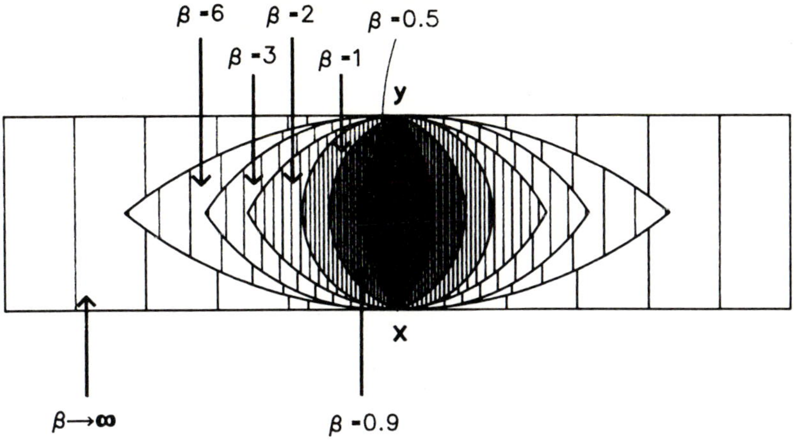

Figure 3a (from [KR])
Lune-based neighbourhoods N(x,y,β)
for various β>0

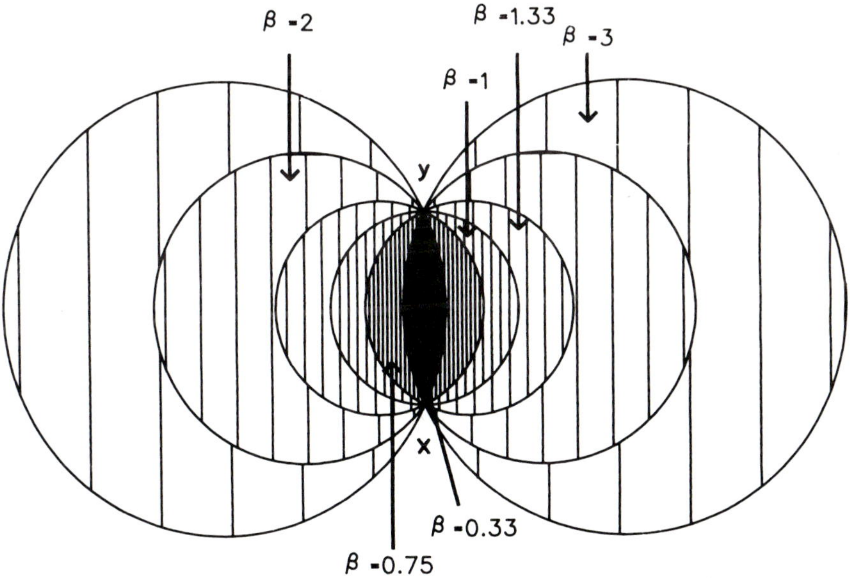

Figure 3b (from [KR])
Circle-based neighbourhoods n(x,y,β),
for various β>0

β - skeletons are powerful morphological decompositions which allow the construction and visualization of a spectrum of internal shape at various densities and scales. The skeleton structures can serve as structural benchmarks, similar to members of the discrete family of graphs discussed earlier, but possessing the sensitivity necessary to detect subsets of clusters or patterns within a point set. The morphological characteristics of these theoretically generated benchmarks can be correlated with those measured in empirical point sets to enhance their description and better hypothesize what their underlying generative processes might be. Figure 4 illustrates a sequence of β - skeletons.

Although Kirkpatrick and Radke [KR] illustrate the application of β - skeletons using line networks, it is clear that their use as descriptors of morphological structure is universal. The notion of β - neighbours can prove useful for detecting the influence of the existing global boundary of a point set on the outlying or "extreme" points of the set. Similarly, they can provide a logic for the generation of a global boundary. Radke [R2] suggests that structure such as the convex hull and α-shapes are often not good boundary generators in real world situations as points on the periphery of a point set lie on the boundary itself. The argument suggests that in biological and ecological studies, a buffer zone is needed between the extreme point locations and the boundary of the point set.

Radke [R2] suggests alternative schemes that generate protective boundaries for a point set. These generation techniques are similar to techniques found in cluster analysis that are often a first step in "unsupervised learning" or detection of a point set's spatial association when "little apriori knowledge" is available. [DB]. The first scheme uses a disc with a constant radius K (whose center is each point in S) to generate a protective buffer around each point in S. Simply put, for all x in a point set S, their neighbourhood N(x,K), represents a bounded buffer zone around each x. The union of all N(x,K) generates the bounded area of point set S. This scheme generates a protective boundary around "extreme" points otherwise not protected by the convex hull or α-shapes. Subsets may result from this scheme since it is dependent on the value of K and the density of the point set itself (See Figure 5).

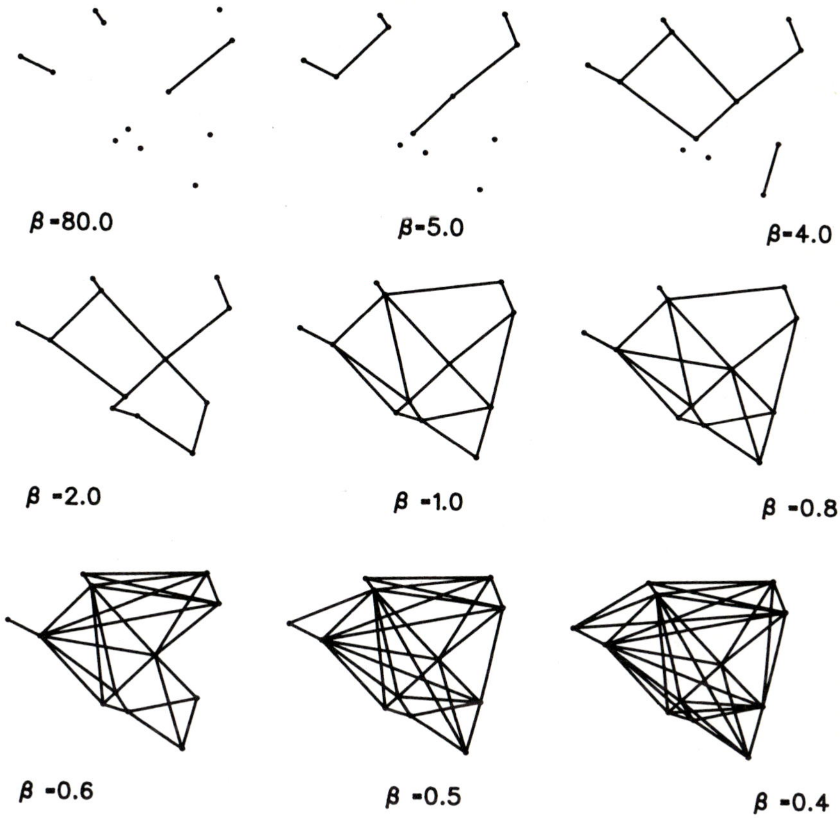

Figure 4 (FROM [KR]) SEQUENCE OF β-SKELETONS ON A FIXED PRINT SET

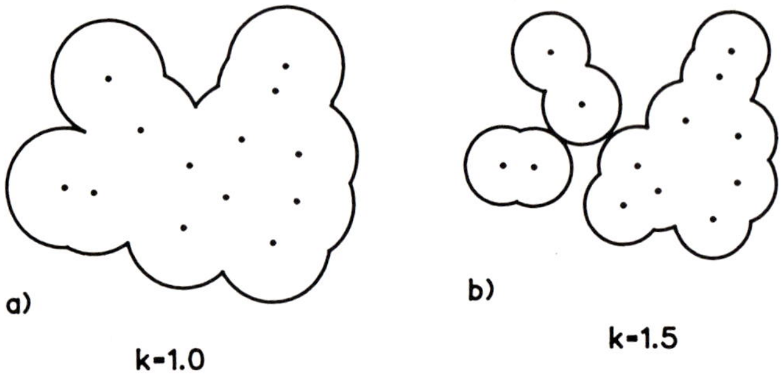

Figure 5 Constant radius neighbourhoods n(x,k)

Although a buffer now exists around the members of the point set, it does not necessarily exert an equal influence or bias on the "extreme" points in the pattern. Intuitively, in many real world situations, the density of members of some phenomenon serves to affect their protective capability (there is often greater safety in numbers). For example, many species find greater protection by clustering to defend against predators and when higher densities are not possible, they attempt to increase their protective boundary. Similarly, isolated points in a market area usually need more protection from intervening opportunities than those in densely packed areas. A boundary generation scheme that addresses the issue of nearness of neighbours or point density often provides a protective boundary. These schemes, many of which are rooted in cluster analysis, often use density measures or nearest neighbour distances in the boundary generation scheme (cf.[GA,DB]).

The constant radius disc method described above can be modified to accommodate variations in neighbour distance. For all x in a point set S, their neighbourhood N(x,K) represents a bounded buffer zone whose centre is x with radius $K=d\tau$. Here d represents the point's nearest neighbour distance and τ is a parameter which can be varied to increase or decrease the boundary edge. The union of all N(x,K) generates the bounded area of point set S. The protective boundary around "extreme" points in this instance reflects the impact of nearest neighbours. Similar to the constant radius scheme, subsets may result with a small value of K or when reflexive nearest neighbours exist (See Figure 6). These protective boundaries are also known as a sphere-of-influence hull (cf.[AH, T4]).

Points can be clustered using various schemes. Some schemes similar to the methods just discussed use spatial association as the criterion for clustering, while other schemes may use attributes as decision criteria. When attributes are used to connect and define the point set, spatially isolated points of groups of points often occur in the defined set. Radke [R2] suggests a method of boundary generation for connected point sets which addresses the issue of altering buffer zones to reflect local spatial conditions better. The scheme uses the neighbourhood search spaces of some of the existing members of

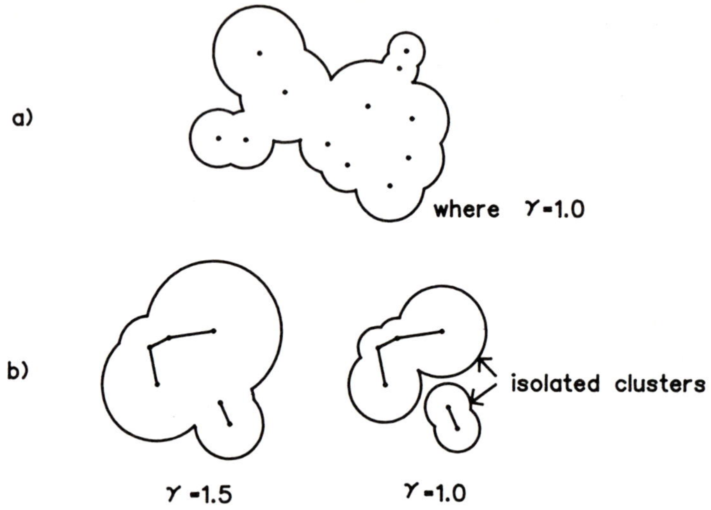

Figure 6 Nearest Neighbour neighbourhoods n(x,k), where k= γd

the family of discrete graphs discussed above.

An obvious expansion of the discrete neighbourhood method is the use of the spectrum of β-neighbourhoods. These neighbourhoods can be used to generate buffer zones with a bounding edges that serve to protect all points in S as well as the connecting corridors between clusters (those connections being generated as a result of attribute characteristics). For all x in a point set S, their neighbourhood $N(x,\beta)$, represents a bounded buffer zone whose centre is x with radius $d(\beta/2)$. Here d represents the points' nearest neighbour distance and β is a parameter which can be varied to increase or decrease the boundary edge. The notion of all $N(x,\beta)$ generates the bounded area of spatial subset clusters in point set S. In addition, bounded buffer zones are generated to protect the connected corridors between clusters. The connection between two clusters is identified by selecting the two nearest neighbours, x and y, each from neighbouring subsets or isolated clusters. The neighbourhood $N(x,y,\beta)$, is defined earlier, is generated for the two neighbouring subsets and is added to the union of all $N(x,\beta)$ to form the buffer zone to protect the points in S. When $\beta=1$, $N(x,y,\beta)$ is the Gabriel Graph "circle of influence" which acts as the buffer zone protecting the connected corridor between clusters (See Figure 7).

Just as the β-neighbourhood was used to generate a spectrum of internal β-skeletons, it can be used here to generate a spectrum of boundaries which protect a point set. However, real world point sets often have superimposed boundaries, such as political boundaries, which do not serve to protect peripheral points equally. The technique employed to measure real world line networks with β-skeletons (cf.[R1,KR]) can be adopted here and β-neighbourhoods can be used to measure the influence or bias of existing boundaries on real world point sets. Figure 8 illustrates a common situation where the state boundary has been used to report agricultural statistics, in this instance, on sheep. However, if we observe the distribution of sheep in Tennessee, the state boundary is not realistic. Employing the β-neighbourhood scheme, a relative boundary influence on the real distribution can be detected. When $\beta < 1$ the real boundary is likely to be outside the state boundary, and when β is very large, the state boundary has very little locational impact on

Figure 7 β- NEIGHBOURHOOD BUFFER ZONE

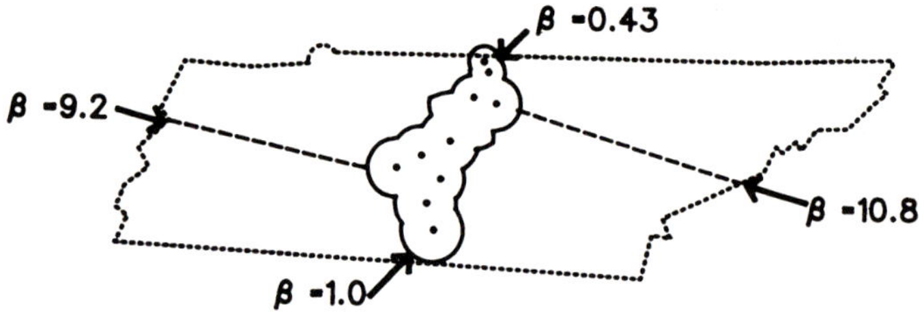

Figure 8 DISTRIBUTION OF SHEEP AND LAMBS (1 DOT=10,000 HEAD)

the distribution, but may have a damaging impact on statistical reporting where density is involved.

Many of the techniques discussed above allow detection of internal, neighbourhood spatial structure revealing subset patterns in point sets that previously have gone undetected. These point sets can then be decomposed and their subsets represented in the abstract as a polygon. Just as β-neighbourhoods address the spatial relationships between points at a local level, neighborhoods can be constructed that reveal the spatial association between subsets, or clusters of points in S. Although some attempts have been made to look at the shape of polygons themselves (cf.[GB]), what is needed is more research into the spatial relationships between polygons.

A continuous family of neighbourhoods can be developed similar to those generated by Kirkpatrick and Radke [KR] which can be used to compare spatial relationships of subsets or clusters of points within a point set S. In the real world, these clusters are often decomposed to a set of polygons, their polygonal edges representing an outer boundary of the subset. The generation of such boundaries can result from a number of logical processes. Some of these are spatially motivated and similar to those discussed above, while others are based on the attribute homogeneity of their member points (cf.[GK]). For a given set of polygons Sp, two polygons Px and Py are β-neighbours if $N(Px,Py,\beta)$ contains no part of a polygon of Sp, other than Px or Py, in its interior.

An internal based structure can be generated, similar to the β-skeleton of a point set S, by joining β-neighbours in Sp. However, the resultant structures can either be line based or polygon based depending on their eventual use. At some small value of β the entire set Sp becomes one cluster and is bounded as one polygon. At some large value of β each cluster or subset remains separate from other subsets in Sp. Figure 9 represents polygon based β-neighbours where a circle based neighbourhood was used and β=1.

When clustering is a function of spatial association, it is probable that cases exist where β-neighbourhoods of a set Sp will approximate the α-shape of the same set of points S. However,

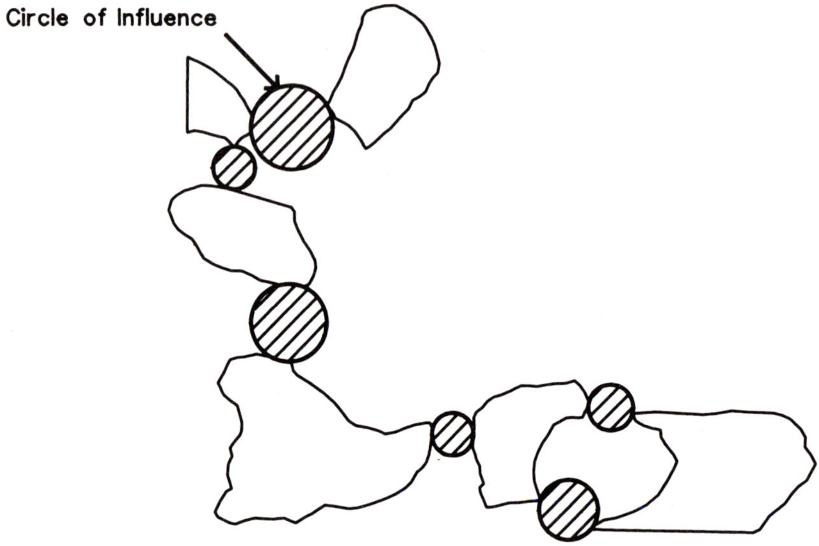

Figure 9 Cluster of Polygons N(Px,Py,β), β=1

the edge or boundary that makes up Sp will only be similar if the generative process for clustering in Sp is the convex hull or α-shape.

The concept of clustering polygons is useful in that it provides a logic for generating a spectrum of shapes and subsets that are similar to α-shapes that can be used as bench marks in recognition problems. A more practical benefit arises when we use the technique in land use planning.

In land use planning a common quest is to locate sites suitable for a given use. This is accomplished by overlaying a number of polygons that contain specific information about attribute characteristics of the site. For example, polygons representing clusters of sampled points where tests of features such as geology, soils, vegetation, slope, etc. have taken place. The planner extracts those intersections from the set of overlaid polygons that are most appropriate for the land use being planned for. These intersections, that are also represented as polygons, are often small in size as they represent common areas of disaggregated attribute information. The planner on the other hand is often looking for large contiguous areas where economies of land use can be realized. As a result, many of the smaller yet appropriate intersection polygons are not considered. If grouped into clusters, many of these small but appropriate polygons could make up a suitable site and satisfy the planning criteria. The β-neighbourhood method can provide the logic for this spatial clustering.

7. Discussion

This paper has examined several methods of detecting spatial structure in a point set. It is clear that those common to spatial analysis are for the most part summarizations, and provide little insight into pattern variation within a point set. In contrast, recent advances in computational morphology appear to have overcome many of the constraints uncovered in spatial analysis. In particular, α-shapes and β-skeletons have provided generalizations which can decompose point sets into meaningful structures. These decompositions help detect patterns within the point sets that often may have previously gone undetected.

It is clear that many extensions of α-shapes and β-skeletons exist and will prove useful in the study of point patterns. After examination :n this paper, β-neighbourhoods conceptually have extensions to higher dimension sets where edges and polygons become the focus of attention. Continued interaction between computational morphology and spatial analysis promises to solve a variety of problems.

This paper set out to show that decomposition techniques appear more promising for recognizing morphologic structure than those now being used in spatial analysis. This will undoubtedly prompt more rigorous experimentation on "real world" data sets and bring us closer to an understanding of form.

REFERENCES

[AH] Avis, D. and J. Horton, "Remarks on the Sphere of Influence Graph," Annals of the New York Academy of Sciences, 440, 1985, pp. 323-327.

[AT] Akl, S.G. and G.T. Toussaint, "Efficient convex hull algorithms for pattern recognition applications," Proceedings of the Fourth International Joint Conference on Pattern Recognition, Kyoto, Japan, 1978, pp.1-5.

[CE] Clark, P.J. and F.C. Evans, "Distance to nearest neighbour as a measure of spatial relationship in population," Ecology 35, October 1954, pp.445-453.

[Bo] Boots, B.N., "Weighing Theissen polygons," Economic Geography 56 (3) July 1980, pp. 248-259

[BG] Boots, B.N. and A. Getis, "Probability model approach to map pattern analysis," Progress in Human Geography 1 (2), 1977, pp. 264-287.

[CO] Cliff, A.D. and J.K. Ord, Spatial Processes: Models and Applications, Pion, London, 1981.

[Da] Dacey, M.F., "The spacing of river towns," Annals of the Association of American Geographers 50, 1960, pp.59-61.

[DB] Davies, D.L. and D.W. Bouldin, "A cluster separation measure," IEEE Trans. on Pattern Analysis and Machine Intelligence PAMI-1 (2), April 1979, pp. 224-227.

[EKS] Edelsbrunner, E, D.G. Kirkpatrick and R. Seidel, "On the shape of a set of points in the plane," IEEE Trans. Information Theory 29 (4), July 1983, pp. 551-559.

[FS] Freeman, H. and R. Shapira, "Determining the minimum-area encasing rectangle for an arbitrary closed curve," Comm ACM 18, July 1975, pp. 409-413.

[GB] Getis, A. and B.N. Boots, Models and Spatial Process, Cambridge University Press, Cambridge, 1978.

[GK] Gowda, K.C. and G. Krishna, "Agglomerative clustering using the concept of mutual nearest neighbourhood," Pattern Recognition 10, 1978, pp. 105-112.

[GA] Griffith, D. A. and C.G. Amrhein, "An evaluation of correction techniques for boundary effects in spatial statistical analysis: Traditional methods," Geographical Analysis, 15 (4), October 1983, pp. 352-360.

[HCF] Haggett, P., A.D. Cliff and A. Frey, Locational Analysis in Human Geography, Arnold, London 1977.

[JBD] Johnston, B. T. Bailey and R. Dubes, "A variation on a nonparametric clustering method," IEEE Trans. on Pattern Analysis and Machine Intelligence PAMI-1 (4), October 1979, pp. 400-408.

[KR] Kirkpatrick, D.B. and J.D. Radke, "A framework for computational morphology,' in Computational Geometry, G.T. Toussaint, ed., Elsevier Science Publishers B.V. (North Holland), 1985, pp. 217-248.

[Le] Lefever, D.W., "Measuring geographic concentration by means of the Standard Deviational Ellipse," The American Journal of Sociology 32 (1), January 1926, pp. 88-94.

[MEP] Mardia, K.V., R. Edwards and M.L. Puri, "Analysis of central place theory," Bulletin of the International Statistical Institute, 47 (2), 1977, pp. 93-110.

[MS] Matula, D.W. and R.R. Sokal, "Properties of Gabriel graphs relevant to geographic variation research and the clustering of points in the plane," Geographical Analysis, 12, pp. 205-222.

[P] Pavlidis, T., Structural Pattern Recognition, Springer-Verlag, New York, 1977.

[Pa] Pavel, M., "Algebraic, topological and categorical aspects of pattern recognition: a survey," Pattern Recognition, 14 (1-6), 1981, pp. 117-120.

[PW] Pinder, D.A. and M.E. Witherick, "A Modification of Nearest-Neighbour Analysis for use in Linear Situations," Geography 60 (1-266), January 1975, pp.16-23.

[R1] Radke, J., "Pattern recognition in circuit networks," Ph.D thesis, Department of Geography, University of British Columbia, November 1982.

[R2] Radke, J., "On measuring the influence of boundaries," Presented at the annual Canadian Association of

Geographer, Nanaimo, B.C. May 1984.

[Ri] Ripley, B., Spatial Statistics, Wiley, New York, 1981.

[SJ] Smith, S.P. and A.K. Jain, "Testing for uniformity in multidimensional data," IEEE Trans. on Pattern Analysis and Machine Intelligence PAMI-6 (1), January 1984, pp. 73-80.

[Su] Supowit, K.J., "The relative neighbourhood graph, with an application to minimum spanning trees," J. ACM 30(3), July 1983, pp.428-448.

[Ta] Taylor, P.J., Quantitative Methods in Geography, Houghton Mifflin Company, Boston, 1977.

[T1] Toussaint, G.T., "The convex hull as a tool in pattern recognition," Proceedings AFOSR Workshop in Communication Theory and Applications, Provincetown, MA, September 1978, pp.43-46.

[T2] Toussaint, G.T., "The relative neighbourhood graph of a finite planar set," Pattern Recognition 12 (4), 1980, pp. 261-268.

[T3] Toussaint, G.T., "Pattern recognition and geometrical complexity," Proceeding of the Fifth International Conference on Pattern Recognition, December 1980, pp. 1324-1347.

[T4] Toussaint, G.T., "A graph-theoretical Primal Sketch," Technical Report N. SOCS-86.17, School of Computer Science, McGill University, Montreal, October 1986

[TB] Toussaint, G.T. and B.K. Bhattacharya, "On geometric algorithms that use the furthest-point Voronoi diagram," Technical Report No. socs-81.3, School of Computer Science, McGill University, Jan. 1981.

[TM] Toussaint, G.T. and R. Menard, "Fast algorithms for computing the planar relative neighbourhood graph," Proc. Fifth Symposium on Operations Research, Koln, August 1980, pp. 425-428.

[U] Urquhart, R., "Graph theoretical clustering based on limited neighbourhood sets," Pattern Recognition, 15 (3), 1982, pp. 173-187.

[Z] Zahn, C.T., "Graph-theoretical methods for detecting and describing gestalt clusters," IEEE Trans. Computer C-20, January 1971, pp.68-86.

Computational Morphology
G.T. Toussaint (editor)

ORTHO-CONVEXITY AND ITS GENERALIZATIONS

Gregory J. E. Rawlins

Department of Computer Science, University of Indiana
Bloomington, Indiana, U.S.A.

Derick Wood

Department of Computer Science, University of Waterloo
Waterloo, Ontario, Canada

In this paper we report on one aspect of work in progress in restricted-orientation geometry. We investigate various alternative definitions of convexity in this setting rather than the standard definition. At the same time, we indicate how a firm mathematical foundation for these notions can be obtained.

1. Introduction

The notion of a *convex set* has a long and interesting history. Fenchel in his survey paper, Fenchel(1983), traces convexity to a form of proto-convexity in the works of Archimedes! However, it is not until quite recently (early twentieth century) that convexity as a unifying notion emerged and found a wide application in the works of Brunn, Minkowski, and others. Still more recently, the young field of computational geometry became interested in and devised algorithms for the computation of convex hulls. But, it was only recently that computational geometry began to consider definitions of convexity deemed more appropriate to the field.

In several application areas of computational geometry the only lines of interest are *orthogonal lines* or *ortho-lines*, that is, lines parallel to the x- or y-coordinate axes. This gives rise, in a natural way, to ortho-polygons (polygons whose edges are orthogonal line segments) and so — inevitably — to the question: What is an appropriate definition of a convex set in this setting?

Our starting point is the work of Montuno and Fournier (1982), Nicholl, Lee, Liao, and Wong (1983), and Ottmann, Soisalon-Soininen, and Wood (1984). These three groups tried to define an appropriate notion of a "convex hull" for ortho-polygons and point sets in the plane. Jumping ahead somewhat, they thought it reasonable that the convex hull be forced to be an ortho-polygon as well. This requirement, although a reasonable one, is somewhat restrictive, (and unnecessary) as we shall see. It also implies that convex polygons and sets have also to be defined and studied with this requirement in view. All three papers failed to cover the mathematical foundations of their proposals in sufficient depth. They each introduced the following definition, which is one possible definition of an *"ortho-convex" polygon*:

Definition 1.1

An ortho-polygon **P** is *ortho-convex* if the intersection of **P** with any horizontal or vertical line is empty, a point, or a line segment.

The ortho-polygons of Figures 1.1(a) and (b) are both ortho-convex, while that of Figure 1.1(c) is not.

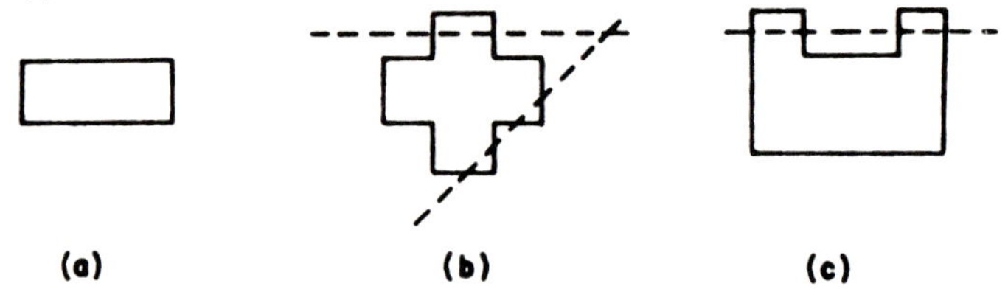

Figure 1.1: Two ortho-convex ortho-polygons and one non-ortho-convex ortho-polygon.

This definition should be compared with the traditional one of a convex set, namely:

Definition 1.2

A polygon **P** is *convex* if the intersection of **P** with any line is empty, a point, or a line segment.

The polygon of Figure 1.1(a) is convex in this sense, those of Figures 1.1(b) and (c) are not.

In Rawlins and Wood (1987) these ideas are placed in a more general framework. They observe that *arbitrary* polygons can be "ortho-convex", if the definition is simplified to:

Definition 1.3

A polygon **P** is *ortho-convex* if the intersection of **P** with any horizontal or vertical line is empty, a point, or a line segment.

Under this definition the polygon of Figure 1.2 is ortho-convex. These examples illustrate the general inferences that (i) a convex polygon is always ortho-convex and, (ii) there are ortho-convex polygons that are not convex.

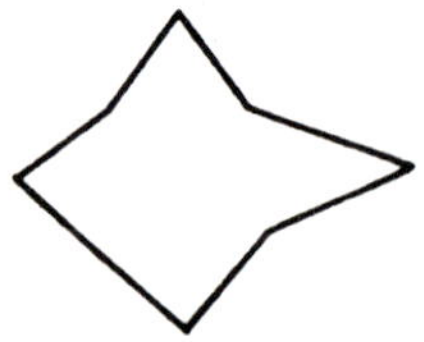

Figure 1.2: An ortho-convex polygon that is not an ortho-polygon.

These four papers are the starting point for our investigation of convexity. We address three basic issues. First, is this the only reasonable definition of convexity? Second, which characterization theorems for convex sets can be carried over to ortho-convex sets? Third, what is the appropriate definition of the ortho-convex closure, that is, the ortho-convex hull?

In the remainder of this paper we discuss these issues. In Section 2 we consider characterization theorems, in Section 3 we consider a second approach to ortho-convexity, and in Section 4 we consider convex hulls. In Section 5 we discuss restricted-orientation geometry, in Section 6 we introduce the unifying framework of convexity spaces, and in Section 7 we summarize our efforts and indicate some future directions.

2. Characterizations of ortho-convexity

There are a number of different and useful characterizations of convexity. These include, where **P** is a planar convex polygon:

(1) **P** is the intersection of all convex polygons which contain it.

(2) For all points p, $p \notin \mathbf{P}$ if and only if there is a line separating p from **P**.

(3) **P** is the intersection of all halfplanes which contain it.

(4) For all points p and q, $p, q \in \mathbf{P}$ if and only if the line segment joining them is in **P**.

In this section we introduce corresponding characterization theorems for ortho-convex polygons. Let **P** be a simple planar polygon throughout. One basic result that we make much use of is that an ortho-convex polygon has exactly four extremal edges. An edge is a *W-edge* if its normal points to the west. N-, S-, and E-edges are defined similarly. We say an edge is *west-extremal* if it is a W-edge and the two edges adjacent to it lie to its east We define *north-*, *south-*, and *east-extremal*, similarly, and say an edge is *extremal* if it is z-extremal, for some z in {north, south, east, west}.

Lemma 2.1

Let **P** be a simple ortho-convex polygon. Then **P** has one z-extremal edge, for each z in {north, south, east, west}.

Proof: At least one of each kind of extremal edge must exist, since **P** is a polygon and, hence, is bounded. If **P** has two west-extremal edges, say, then the ortho-convexity of **P** is violated. This is because there is an ortho-line which intersects **P** in two or more disconnected line segments, namely, the ortho-line passing through the easternmost west-extremal edge also passes through **P** elsewhere. □

Lemma 2.2

Let $I(\mathbf{P})$ be the intersection of all ortho-convex polygons containing **P**. Then $I(\mathbf{P})$ is ortho-convex.

Proof: Let p and q be in $I(\mathbf{P})$ such that p, q lie on a horizontal line (or on a vertical line). If no such points exist, then $I(\mathbf{P})$ is trivially ortho-convex. Clearly, for each ortho-convex polygon **Q** containing **P**, p and q are both in **Q**. Therefore, the line segment joining p and q is in each such **Q** and, hence, in $I(\mathbf{P})$. Therefore, $I(\mathbf{P})$ is ortho-convex. □

Theorem 2.3

Let **P** be a polygon. **P** is ortho-convex if and only if $\mathbf{P}=I(\mathbf{P})$.

Proof: If $\mathbf{P}=I(\mathbf{P})$, then **P** is ortho-convex, by Lemma 2.2. Conversely, if **P** is ortho-convex, then not only is $\mathbf{P} \subseteq I(\mathbf{P})$, but also **P** is an ortho-convex polygon that contains **P**; therefore, $I(\mathbf{P}) = \mathbf{P}$. □

We generalize Theorem 2.3 somewhat by:

Theorem 2.4

Let **P** be a polygon and $OCS(\mathbf{P})$ be the intersection of all ortho-convex sets containing **P**. Then **P** is ortho-convex if and only if $\mathbf{P} = OCS(\mathbf{P})$.

Proof: That $OCS(\mathbf{P})$ is ortho-convex follows in a similar manner to the proof of Lemma 2.2, while the theorem statement follows in a similar manner to the proof of Theorem 2.3. □

Definition 2.3

A curve is *x-monotone* (*y-monotone*) if its intersection with any vertical line (horizontal line) is empty, a point, or a line segment. A *stairsegment* is a finite curve that is both x- and y-monotone. A *staircase* is a stairsegment made up of alternating horizontal and vertical line segments.

See Figure 2.1 for four examples of staircases. To obtain our next characterization theorem we first characterize the appearance of an ortho-convex polygon in terms of staircases.

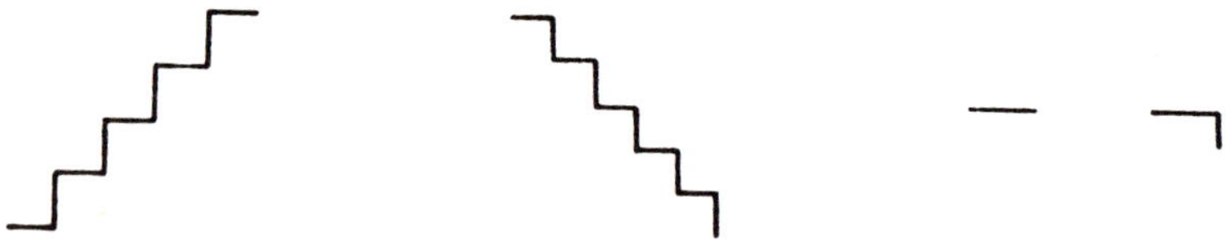

Figure 2.1: Four staircases.

Definition 2.4

An *ortho-ray* is a horizontal or vertical ray. A *stairline* is a curve that is both x- and y-monotone. An *ortho-stairline* is a stairline made up of two ortho-rays connected by a staircase; see Figure 2.2 for two examples.

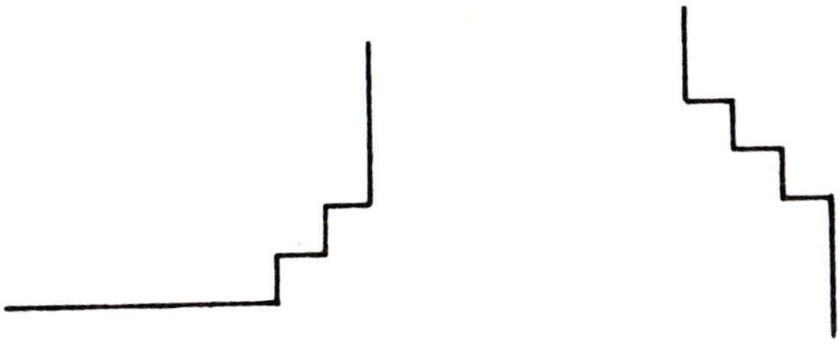

Figure 2.2: Two ortho-stairlines.

When **P** is an ortho-convex polygon each horizontal or vertical line, that passes through its interior, cuts its boundary at two points. Thus, the following characterization of the boundary of **P** should come as no surprise. We first need:

Definition 2.5

Given a polygon **P** we say that an a-extremal edge and a b-extremal edge are *adjacent* if b follows a in the list "north, east, south, west", where "north" follows "west".

Theorem 2.5

Let **P** be an ortho-convex polygon. Then the boundary of **P** is made up of four stairsegments. Further, if **P** is also an ortho-polygon, then the boundary consists of four staircases.

Proof: **P** has four extremal edges. The boundary between any two adjacent extremal edges together with one of the extremal edges is both x- and y-monotone (otherwise, **P** is not ortho-convex). Clearly there are exactly four such stairsegments. This completes the proof of the first statement, the second follows because each stairsegment is made up of alternating horizontal and vertical line segments. □

Theorem 2.6

Let **P** be a polygon. **P** is ortho-convex if and only if for all points $p \notin$ **P**, there is a stairline separating p from **P**.

Proof: If, for all points $p \notin$ **P**, there is a stairline separating p from **P**, then **P** contains neither holes nor notches; see Figure 2.3. Conversely, if **P** is ortho-convex and $p \notin$ **P**, then we can easily construct a stairline separating them; see Figure 2.4. □

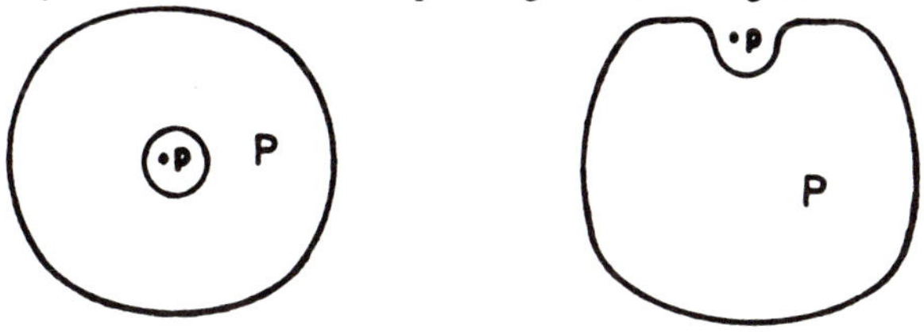

Figure 2.3: A separable polygon cannot contain holes or notches.

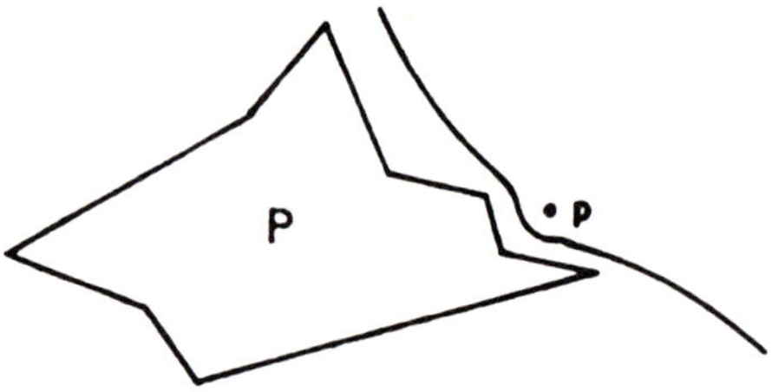

Figure 2.4: A point not in an ortho-convex polygon can always be separated.

Definition 2.6

A *three-quarter plane* is the complement of an open quadrant. They come in four varieties; see Figure 2.5.

Figure 2.5: The four three-quarter planes.

Lemma 2.7

Let **P** be a polygon and let $TQ\,(\mathbf{P})$ be the intersection of all three-quarter planes containing **P**. Then $TQ\,(\mathbf{P})$ is ortho-convex.

Proof: This is similar to the proof of Lemma 2.2 and is omitted. □

Theorem 2.8

Let **P** be a polygon. **P** is ortho-convex if and only if $\mathbf{P} = TQ\,(\mathbf{P})$.

Proof: If $\mathbf{P} = TQ\,(\mathbf{P})$ we immediately infer that **P** is also ortho-convex from Lemma 2.7. Conversely, if **P** is ortho-convex, we need to prove that $TQ\,(\mathbf{P}) = \mathbf{P}$. Note that although $\mathbf{P} \subseteq TQ\,(\mathbf{P})$, **P** is not a three-quarter plane, and, therefore, we cannot conclude our argument as simply as in Theorem 2.3. Let $p \notin \mathbf{P}$; then from Theorem 2.6 there exists a stairline separating p and **P**; let this be S. Also, assume that p is to the north-west of **P**; see Figure 2.4.

Consider a horizontal line H, through p. Since S is a stairline it is x-monotone and therefore H intersects S at exactly one point q, say. Now move down S from q until a new point q' on S is met such that q and q' have different y-coordinates; see Figure 2.6. This must occur because S separates p and **P**. Now q' defines a three-quarter plane that contains **P** but not p. Therefore, $p \notin TQ\,(\mathbf{P})$. In other words, we have shown that any point that is not in **P** is also not in $TQ\,(\mathbf{P})$, that is, $\mathbf{P} = TQ\,(\mathbf{P})$. □

The importance of this result is the demonstration that the ortho-convex hull of an ortho-polygon **P**, see Section 4, is orthogonal.

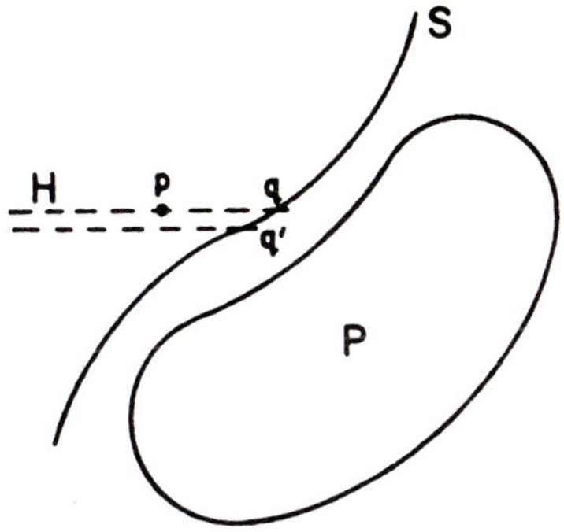

Figure 2.6: A three-quarter plane that contains **P** but not p.

Theorem 2.9

Let **P** be a polygon. **P** is ortho-convex if and only if for all points $p, q \in \mathbf{P}$, that have the same x-coordinate or the same y-coordinate, the line segment (p, q) joining them is in **P**.

Proof: Consider a vertical line V through **P**, that is $V \cap \mathbf{P} \neq \varnothing$. Since **P** is closed and bounded, there is a highest point p and a lowest point s, in $V \cap \mathbf{P}$. (p, s) must be a line segment in **P**, since r and s are in **P** and have the same x-coordinate.

Conversely, if **P** is ortho-convex and $p, q \in \mathbf{P}$ such that p and q have the same x-coordinate, then the vertical line through p and q has a nonempty intersection with **P**. From the definition of ortho-convexity this intersection is a line segment that lies in **P** and contains p and q. But this implies that the line segment (p, q) is in **P** as desired. □

In this case we have another, more general, version of the theorem. But, before stating and proving it we need the following definition.

Definition 2.7

Let p and q be two points in the plane; they define a unique orthogonal rectangle, see Figure 2.7. If they are on the same ortho-line they define a unique orthogonal line segment, the degenerate case of a rectangle. The upper boundary of the rectangle from p to q is called the *upper arm* and the lower boundary is called the *lower arm*.

Figure 2.7: p and q define a unique rectangle.

Theorem 2.10

Let **P** be a polygon. **P** is ortho-convex if and only if, for all points $p, q \in \mathbf{P}$, there is a stairsegment joining them that is in **P**.

Proof: Let **P** be ortho-convex with $p, q \in \mathbf{P}$. If $p = q$, the implication is trivial, therefore, assume $p \neq q$. If p and q are on the same horizontal or vertical line, the implication follows because **P** is ortho-convex. For the remaining case, assume, without loss of generality, that p is to the left of and below q. We show that there is a stairsegment, joining p and q, wholly in **P**, as follows. First, if one of the upper or lower arms joining p and q is wholly in **P**, this is the required stairsegment. Otherwise, the lower arm intersects the boundary of **P** at points p' and q', say. We claim that the curve formed from the edge (p, p'), together with the boundary of **P** from p' to q', and the edge (q', q) is, indeed, a stairsegment; see Figure 2.8.

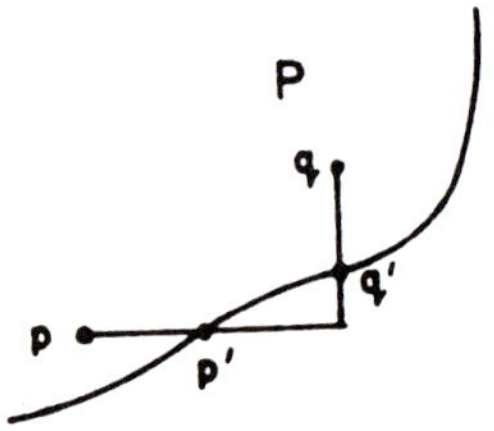

Figure 2.8: A stairsegment wholly in **P** joining p and q.

But this follows immediately from the observation that the bottommost and rightmost extremal edges of **P** cannot be within the rectangle formed by p and q (otherwise the ortho-convexity of **P** would be violated).

Conversely, if there is a stairsegment, wholly in **P**, joining every pair of points p, q in **P**, then this clearly holds for those points p and q on the same horizontal or vertical line. Therefore **P** is ortho-convex. □

3. A second definition of orthogonal convexity

Preparata (private communication) suggested the following definition of ortho-convexity, which we call *box convexity*.

Definition 3.1

A polygon **P** is *box convex* if, for all pairs of points p, q in **P** the orthogonal rectangle defined by them lies wholly in **P**. If $p = q$, then the rectangle defined by them is the point p and if p and q both lie on an ortho-line, then the defined rectangle is the line segment joining them.

Each pair of points in the plane defines a unique orthogonal rectangle when the points are taken as the endpoints of one of the diagonals. Under this definition every orthogonal rectangle is box convex and no other polygon is! This is easy to see. Let **P** be a polygon that is not a rectangle. If it has at least one edge that is neither horizontal nor vertical, two distinct points p and q on it define a rectangle not wholly in **P**; see Figure 3.1. On the other hand, if all edges are horizontal and vertical and there is one reflex vertex, consider two distinct points p and q on different edges adjacent to this vertex; see Figure 3.2. Clearly, the rectangle they define does not lie wholly in **P**.

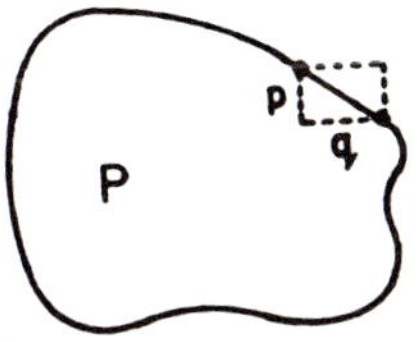

Figure 3.1: A rectangle, defined by a non-ortho-edge, not wholly in **P**.

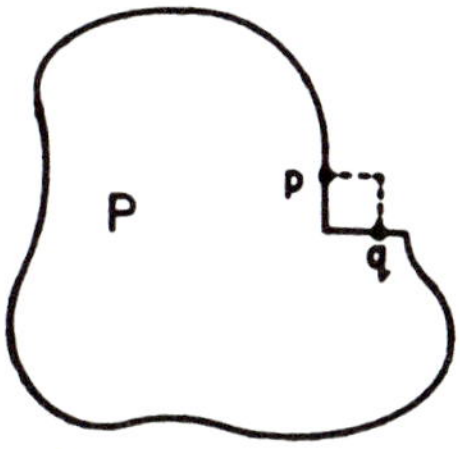

Figure 3.2: A rectangle, defined by a reflex vertex, not wholly in **P**.

Because this notion of convexity is very restrictive, one may wish to consider weaker, but similar notions. We can relax this to:

Definition 3.2

A polygon **P** is *triangle convex* if, for all pairs of points p, q in **P**, one of the triangles defined by them lies wholly in **P**. A *triangle* is the diagonal from p to q together with either the lower or upper arms of the defined rectangle; see Figure 3.3.

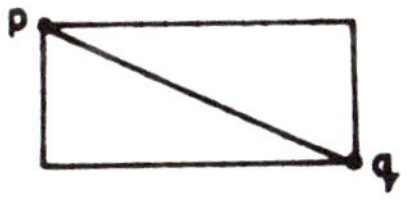

Figure 3.3: The two triangles defined by p and q.

However, triangle convexity is, *for ortho-polygons*, equivalent to box convexity: observe in Figure 3.1 that neither triangle lies wholly in **P**.

Therefore another approach might be to consider only the boundary of the defined rectangle; in particular, the boundaries of the upper or lower triangles excluding the diagonal, that is, the *arms*.

Definition 3.3

A polygon **P** is *arm convex* if, for all pairs of points p, q in **P**, one of the arms of the defined rectangle lies wholly in **P**.

In this case we do obtain more convex polygons, since the symmetric cross in Figure 3.4(a) is arm convex; however, the asymmetric cross in Figure 3.4(b) is not.

Figure 3.4: An arm-convex polygon and a non-arm-convex polygon.

Since an arm is a staircase of length 2, a *2-staircase*, we could extend this notion to *m-staircase convexity* (we assume that every k-staircase, $1 \leq k < m$, is also an m-staircase). Do these notions of convexity display the characteristics of the usual convexity? A general theory of convexity helps us to answer such questions.

4. Ortho-convex hulls

Definition 4.1

Let **P** be a polygon. Then the *convex hull* of **P** is the smallest convex polygon containing **P**. We denote it by *CH* (**P**).

It can be proved that the convex hull is well-defined and unique. When given a non-convex polygon **P** it is often more convenient to deal with its convex hull, rather than directly with **P** itself. The same observation holds for non-ortho-convex polygons, the only difference being how should an ortho-convex hull be defined? Do we require the resulting polygon to be orthogonal? Even if we do, then there appears to be more than one reasonable definition of an ortho-convex hull. Without more ado let us give the first, obvious, definition.

Definition 4.1

Let **P** be a polygon. Then its *ortho-convex ortho-hull* is the smallest ortho-convex ortho-polygon containing **P**. We denote it by *OCOH* (**P**).

Note that we require the resulting polygon to be both ortho-convex *and* orthogonal!

Theorem 4.1

Let **P** be an orthogonal polygon. Then *OCOH* (**P**) is unique and, therefore, well-defined. Also *OCOH* (**P**) = *I* (**P**) = *TQ* (**P**) = *OCS* (**P**).

Proof: The first statement follows by way of contradiction. If there are two smallest ortho-convex polygons containing **P**, then they must be incomparable. But this implies their intersection is smaller, it contains **P**, and it is ortho-convex — a contradiction.

The second statement follows, essentially, by demonstrating that *OCS* (**P**) = *OCOH* (**P**). Now, because **P** $\subseteq$ *OCOH* (**P**), *OCS* (**P**) $\subseteq$ *OCS* (*OCOH* (**P**)) and, because *OCOH* (**P**) $\subset$ *OCS* (**P**), *OCS* (*OCOH* (**P**)) $\subset$ *OCS* (*OCS* (**P**)) = *OCS* (**P**). Therefore

$OCS\,(OCOH\,(\mathbf{P})) = OCS\,(\mathbf{P})$, but by Theorem 2.4, $OCS\,(OCOH\,(\mathbf{P})) = OCOH\,(\mathbf{P})$. □

We now turn to a simpler definition of ortho-convex hull for which we do not require the resulting polygon to be orthogonal.

Definition 4.2

Let **P** be a polygon. Then its *ortho-convex hull* is the smallest ortho-convex polygon containing **P**. We denote it by $OCH\,(\mathbf{P})$.

Theorem 4.2 First Equivalence Theorem

Let **P** be an orthogonal polygon. Then $OCH\,(\mathbf{P})$ is unique and well-defined, and $OCH\,(\mathbf{P}) = OCOH\,(\mathbf{P})$.

Proof: Essentially this has already been proved implicitly in the proof of Theorem 4.1.
□

The above equivalence theorem simplifies the definition and computation of the ortho-convex (ortho-) hull for ortho-polygons. We now consider another, more restrictive, definition for which we require convexity rather than ortho-convexity.

Definition 4.3

Let **P** be a polygon. Then its *convex ortho-hull* is the smallest convex ortho-polygon containing **P**. We denote it by $COH\,(\mathbf{P})$.

Since $COH\,(\mathbf{P})$ is convex and orthogonal it must be an orthogonal rectangle.

Theorem 4.3

Let **P** be an orthogonal polygon. Then $COH\,(\mathbf{P})$ is the rectangle defined by the extremal edges of **P**, that is, it is the *bounding box* of **P**.

Proof: Trivial. □

So we see that $COH\,(\mathbf{P})$ is the same as the "convex hull" of **P**, where convexity is defined as *box convexity*. To close this section we have a relationship theorem and a computational theorem.

Theorem 4.4

Let **P** be an orthogonal polygon. Then $\mathbf{P} \subseteq OCOH\,(\mathbf{P}) = OCH\,(\mathbf{P}) \subseteq CH\,(\mathbf{P}) \subseteq COH\,(\mathbf{P})$ and equality holds only if and only if **P** is an orthogonal rectangle.

Proof: $OCH\,(\mathbf{P})$ is not necessarily convex, hence $OCH\,(\mathbf{P}) \subseteq CH\,(\mathbf{P})$, and $CH\,(\mathbf{P})$ is not necessarily a rectangle, hence $CH\,(\mathbf{P}) \subseteq COH\,(\mathbf{P})$. If **P** is an orthogonal rectangle equality clearly holds, and if $\mathbf{P} = COH\,(\mathbf{P})$, then **P** is an orthogonal rectangle. □

Theorem 4.5

Let **P** be an arbitrary orthogonal polygon with n vertices. Then $OCH\,(\mathbf{P})$ and $COH\,(\mathbf{P})$ can both be computed in $O\,(n)$ time and $O\,(n)$ space and these are time-and-space-optimal in a comparison-based model of computation.

Proof: See Rawlins and Wood (1988a). □

5. Restricted-Orientation Convexity

In the previous four sections we have been concerned with possible definitions of convexity in an orthogonal framework. In the present section we sketch generalizations of ortho-convexity and box convexity to give restricted-oriented convexity in which we single out a class of orientations **O**, that is, some subset of all orientations. In other words, we restrict the number of legal orientations and, at the same time, allow more than the two orthogonal orientations. Indeed, we allow **O** to be all orientations. In the next section, in a more abstract setting, we provide a unified framework for restricted-orientation convexity, ortho-convexity, and box convexity. It is convenient to speak of polygons rather than sets of points; however, the reader should note that *the results hold also for arbitrary connected point sets.* Let us turn to the first generalization.

Definition 5.1

A polygon **P** is **O**-*convex* if its intersection with any **O**-line is empty, a point, or a line segment.

Immediately, we obtain:

Lemma 5.1

(i) All convex polygons are **O**-convex.

(ii) If **C** is a nonempty collection of **O**-convex sets, then $\cap$ **C** is **O**-convex.

Proof: (i) is immediate and (ii) follows by a similar argument to that used in Lemma 2.2. □

We obtain a characterization theorem analogous to Theorems 2.3, 2.6, 2.8, and 2.9.

Theorem 5.2 Characterization Theorem

(i) A polygon **P** is **O**-convex if and only if **P** is the intersection of all **O**-convex sets containing **P**.

(ii) A polygon **P** is **O**-convex if and only if, for all $p \notin$ **P**, there is an **O**-stairline separating p from **P**.

(iii) A polygon **P** is **O**-convex if and only if **P** is the intersection of all **O**-halfplanes containing **P**.

(iv) A polygon **P** is **O**-convex if and only if, for all $p, q \in$ **P**, there is an **O**-stairsegment from p to q lying wholly in **P**.

The notions of **O**-stairlines, **O**-halfplanes, etc. are appropriate generalizations of orthogonal stairlines, three-quarter planes, etc. Having stated the basic results about **O**-convex polygons we turn to the notion of the **O**-hull.

Definition 5.2

Let **P** be a polygon and **O** be a set of orientations. Then the **O**-*hull* of **P** is denoted by **O(P)** and is defined by **O(P)** is the intersection of all **O**-convex polygons containing **P**.

We can now prove:

Theorem 5.3

(i) For each polygon **P**, $\mathbf{O}(\mathbf{P})$ exists and is the smallest **O**-convex set containing **P**.

(ii) A polygon **P** is **O**-convex if and only if $\mathbf{P} = \mathbf{O}(\mathbf{P})$.

Notc that $\mathbf{P} \subseteq \mathbf{O}(\mathbf{P})$ even when $\mathbf{P} = \varnothing$ or $\mathbf{O} = \varnothing$.

Even when a polygon is allowed to have holes, that is, it is not simply connected, its **O**-hull is simply connected, once **O** is nonempty.

Lemmas 5.4 Simple Connectedness

Let **P** be a polygon possibly having holes and $\mathbf{O} \neq \varnothing$; then $\mathbf{O}(\mathbf{P})$ is simply connected.

The next result characterizes when a point that is outside **P** is inside the **O**-hull of **P**.

Theorem 5.5 Separation Theorem

Let **P** be a polygon and $p \notin P$. Then $p \in \mathbf{O}(\mathbf{P})$ if and only if there is a $\theta \in \mathbf{O}$ such that the θ-line through p intersects **P** in at least two points on either side of p.

Proof: See Rawlins and Wood(1988a). □

We define the *x-convex hull* of a polygon **P** to be the intersection of all y-monotone polygons containing **P** and, similarly, for the y-convex hull. Now, in Ottmann, Soisalon-Soininen, and Wood (1984, 1986), it is stated that the ortho-convex hull can be obtained by forming the x-convex hull and then the y-convex hull, or vice versa. This is generalized in Rawlins and Wood (1988a) to finitely many orientations and in Soisalon-Soininen and Wood (1984) a similar result is proved for "closure". Finally, in Preparata and Shamos (1986) it is shown that the ortho-convex hull is the union of the x- and y-convex hulls. We now refine and generalize these results considerably.

Theorem 5.6 Decomposition Theorem

Let **P** be a polygon. Then for all sets of orientations $\mathbf{O}_1$ and $\mathbf{O}_2$,

$$
\begin{aligned}
(\mathbf{O}_1 \cup \mathbf{O}_2)(\mathbf{P}) &= \mathbf{O}_1(\mathbf{O}_2(\mathbf{P})) \\
&= \mathbf{O}_2(\mathbf{O}_1(\mathbf{P})) \\
&= \mathbf{O}_1(\mathbf{P}) \cup \mathbf{O}_2(\mathbf{P})
\end{aligned}
$$

Proof: First, note that as $\mathbf{P} \subseteq \mathbf{O}(\mathbf{P})$, for all **O** and all **P**, we have $\mathbf{O}_1(\mathbf{P}) \subseteq \mathbf{O}_2(\mathbf{O}_1(\mathbf{P}))$ and $\mathbf{O}_2(\mathbf{P}) \subseteq \mathbf{O}_1(\mathbf{O}_2(\mathbf{P}))$. Moreover since $\mathbf{P} \subseteq \mathbf{O}_1(\mathbf{P})$, we obtain $\mathbf{O}_2(\mathbf{P}) \subseteq \mathbf{O}_2(\mathbf{O}_1(\mathbf{P}))$ and, similarly, $\mathbf{O}_1(\mathbf{P}) \subseteq \mathbf{O}_1(\mathbf{O}_2(\mathbf{P}))$. Therefore

$$\mathbf{O}_1(\mathbf{P}) \cup \mathbf{O}_2(\mathbf{P}) \subseteq \mathbf{O}_2(\mathbf{O}_1(\mathbf{P}))$$

and

$$\mathbf{O}_1(\mathbf{P}) \cup \mathbf{O}_2(\mathbf{P}) \subseteq \mathbf{O}_1(\mathbf{O}_2(\mathbf{P}))$$

Second, $\mathbf{O}_1(\mathbf{P}) \subseteq (\mathbf{O}_1 \cup \mathbf{O}_2)(\mathbf{P})$, since $\mathbf{O}_1 \subseteq \mathbf{O}_1 \cup \mathbf{O}_2$ and every $(\mathbf{O}_1 \cup \mathbf{O}_2)$-convex set is $\mathbf{O}_1$-convex (and $\mathbf{O}_2$-convex). Therefore $\mathbf{O}_2(\mathbf{O}_1(\mathbf{P})) \subseteq \mathbf{O}_2((\mathbf{O}_1 \cup \mathbf{O}_2)(\mathbf{P})) = (\mathbf{O}_1 \cup \mathbf{O}_2)(\mathbf{P})$ and, similarly, $\mathbf{O}_1(\mathbf{O}_2(\mathbf{P})) \subseteq (\mathbf{O}_1 \cup \mathbf{O}_2)(\mathbf{P})$.

Third, we demonstrate that

$$(\mathbf{O}_1 \cup \mathbf{O}_2)(\mathbf{P}) \subseteq \mathbf{O}_1(\mathbf{P}) \cup \mathbf{O}_2(\mathbf{P}).$$

If $\mathbf{P}$ is empty or $\mathbf{O}_1 \cup \mathbf{O}_2$ is empty, then the result is trivial. So, we assume that both $\mathbf{P}$ and $\mathbf{O}_1 \cup \mathbf{O}_2$ are nonempty. Consider an arbitrary point p in $(\mathbf{O}_1 \cup \mathbf{O}_2)(\mathbf{P})$. If $p \in P$, then $p \in \mathbf{O}_1(\mathbf{P}) \cup \mathbf{O}_2(\mathbf{P})$. Therefore, assume p is not in $\mathbf{P}$. In this case by Theorem 5.5, there is a θ-line through p, for some $\theta \in \mathbf{O}_1 \cup \mathbf{O}_2$, that intersects $\mathbf{P}$ in at least two points on either side of p. If $\theta \in \mathbf{O}_1$, then $p \in \mathbf{O}_1(\mathbf{P})$, otherwise $p \in \mathbf{O}_2(\mathbf{P})$. In other words $p \in \mathbf{O}_1(\mathbf{P}) \cup \mathbf{O}_2(\mathbf{P})$. □

For the final result of this section we state the generalization of the stairsegment theorem for ortho-convex polygons.

Theorem 5.7 Stairsegment Characterization Theorem

Let $\mathbf{O}$ be a set of orientations and $\mathbf{P}$ be a polygon. Then $\mathbf{P}$ is $\mathbf{O}$-convex if and only if the portions of its boundary between any two consecutive $\mathbf{O}$-extremal points are maximal $\mathbf{O}$-stairsegments.

Proof: See Rawlins and Wood (1988a). □

We now consider the second generalization.

Definition 5.3

Let $\mathbf{P}$ be a polygon and $\mathbf{O}$ be a set of orientations. Then $\mathbf{P}$ is *strongly* $\mathbf{O}$-convex if the intersection of $\mathbf{P}$ with any $\mathbf{O}$-stairline is empty, a point, or an $\mathbf{O}$-stairsegment.

This is a generalization of box convexity as the next theorem makes clear.

Theorem 5.8

Let $\mathbf{P}$ be a polygon and $\mathbf{O}$ be a set of orientations. Then $\mathbf{P}$ is strongly $\mathbf{O}$-convex if and only if $\mathbf{P}$ is a convex $\mathbf{O}$-polygon.

Proof: See Rawlins and Wood(1988b). □

6. Generalized convexity

The key result, obtained for each of the various notions of convex set discussed above, is that the intersection of two "convex" sets is once again a "convex" set. This has led geometers to generalize convexity in the following way.

Definition 6.1

Let A be a set and C be a subset of the power set of A, where $\varnothing, A \in C$. Intuitively, C is a collection of "convex" sets. If, for all subcollections B of C,

$$\bigcap_{X \in B} X \in C$$

then we say that the structure (A, C) is a *convexity space*.

Examples of convexity spaces that we have met are:

(1) $A = \mathbb{R}^2$ and C is the set of all convex sets.

(2) $A = \mathbb{R}^2$ and C is the set of all ortho-convex sets.

(3) $A = \mathbb{R}^2$ and C is the set of all $\mathbf{O}$-convex sets, for some arbitrary $\mathbf{O}$.

(4) $A = \mathbb{R}^2$ and C is the set of all strongly $\mathbf{O}$-convex sets, for some arbitrary $\mathbf{O}$.

Definition 6.2

Let (A, C) be an convexity space. Then, for all $X \subseteq A$, the C*-hull* of X is denoted by $C(X)$ and is defined by

$$C(X) = \bigcap_{\substack{X \subseteq Y \\ Y \in C}} Y$$

The following theorem can now be proved.

Theorem 6.1

Let (A, C) be a convexity space and $X \subseteq A$ be an arbitrary set. Then $C(X)$ is well defined and is equal to the smallest set in C containing X, where smallest is defined in the usual way by means of set inclusion.

7. Future directions

In this paper we have introduced two basic approaches to ortho-convexity and have shown how they may be generalized to give restricted-orientation convexity. Finally, we have indicated that these notions can be unified within the scope of convexity spaces.

This work is just beginning; it leaves many questions unanswered, which we are currently trying to solve.

First, normal convexity can, just as naturally, be defined by way of visibility. That is, a set is convex if and only if all pairs of points are mutually visible (this is a rewording of Property (4) in Section 2). Moreover, if a set is convex there is a point in the set from which all other points are visible (star-shapedness). Is it possible to build a theory of visibility that captures these notions? Some preliminary results are to be found in Rawlins (1987) and Rawlins and Wood (1988b).

Second, is a unifying theory useful computationally? For example, Theorem 5.6 (The Decomposition Theorem) has computational implications for finitely many orientations. Indeed, it is the basis of the $O(n \log n)$ time algorithm for closure in Soisalon-Soininen and Wood (1984). Is there more to be said in this direction?

Third, turning to restricted-orientation geometry, rather than convexity, there are a number of interesting problems. For example, art galleries in which staircases as well as line segments determine visibility, and Boolean masking for restricted-orientation layers of polygons. Some results have already been obtained for these particular problems, but many remain to be solved.

8. References

[1] Fenchel, W., Convexity Through The Ages, in *Convexity and Its Applications*, (P.M. Gruber and J.M. Wills, eds), Birkhäuser Verlag, Boston, 1983.

[2] Montuno, D. Y., and Fournier, A., Finding The X-Y Convex Hull of A Set of X-Y Polygons, Technical Report CSRG-148, University of Toronto, 1982.

[3] Nicholl, T. M., Lee, D. T., Liao, Y. Z., and Wong, C. K., Constructing The X-Y Convex Hull of A Set of X-Y Polygons, *BIT 23*, (1983), 456-471.

[4] Ottmann, Th., Soisalon-Soininen, E., and Wood, D., On The Definition and Computation of Rectilinear Convex Hulls, *Information Sciences 33*, (1984), 157-171.

[5] Ottmann, Th., Soisalon-Soininen, E., and Wood, D., Partitioning and Separating Sets of Orthogonal Polygons, *Information Sciences 42*, (1987), 31-49.

[6] Preparata, F.P., and Shamos, M.I., *Computational Geometry*, Springer-Verlag, New York, 1986.

[7] Rawlins, G.J.E., *Explorations in Restricted-Orientation Geometry*, Doctoral Dissertation, University of Waterloo, 1987.

[8] Rawlins, G.J.E., and Wood, D., On The Optimal Computation of Finitely-Oriented Convex Hulls, *Information and Computation 72*, (1987), 150-166.

[9] Rawlins, G.J.E., and Wood, D., Computational Geometry with Restricted Orientations, *Proceedings of the 13th IFIP conference on System Modelling and Optimization, Springer-Verlag Lecture Notes in Computer Science*, (1988a), to appear.

[10] Rawlins, G.J.E., and Wood, D., Convexity Spaces: A Decomposition Theorem, in preparation, 1988b.

[11] Soisalon-Soininen, E., and Wood, D., Optimal Algorithms to Compute The Closure of A Set of Iso-Rectangles, *Journal of Algorithms 5*, (1984), 199-214.

Computational Morphology
G.T. Toussaint (editor)

Guard Placement in Rectilinear Polygons+

Jörg-Rüdiger Sack

School of Computer Science, Carleton University, Ottawa, Ontario, Canada K1S 5B6

Godfried T. Toussaint

School of Computer Science, McGill University, 805 Sherbrooke St. West, Montréal, Québec, Canada H3A 2K6

Guard placement problems have been extensively studied by mathematicians as well as computer scientists. The classical guard problem posed by Victor Klee is to determine the number of guards sufficient to see any art gallery given as an n-vertex simple polygon. Here we study traditional (or rectilinear) art-galleries, i.e. galleries in which all interior angles are either 270° or 90°. The *Rectilinear Art Gallery Theorem* originally proved by Kahn, Klawe and Kleitman states that any n-vertex rectilinear art gallery can always be guarded by at most $\lfloor n/4 \rfloor$ guards. Here we examine the problem from its computational point-of-view by providing an algorithmic proof of the Rectilinear Art Gallery Theorem. It is demonstrated that guard placement in monotone rectilinear polygons can be done in linear time, while the problem can be solved for arbitrary, n-vertex, rectilinear polygons in O(n loglog n) time.

+ This research was supported by the Natural Sciences and Engineering Council of Canada.

1. INTRODUCTION

1.1 Some Relevant Literature

The following interesting question was raised by Victor Klee: How many guards are always sufficient to guard the interior of an art gallery? The floor-plan of the gallery is given as a simple n -vertex polygon P and guards are to be located at stationary points inside P. The gallery has paintings hanging on walls, as well as sculptures placed inside the gallery; it is thus important not only to guard the walls of the gallery but also its interior. A guard stationed at a point q in P can *see* a point p in P if the line segment pq joining p to q lies inside P (it is assumed that guards can see in all directions at once).

V. Chvátal [4] solved Klee's problem by showing that an arbitrary simple n-vertex polygon can always be guarded using at most $\lfloor n/3 \rfloor$ guards. He also gave an example showing that $\lfloor n/3 \rfloor$ guards may sometimes be necessary. Subsequently Fisk [8] gave a much simpler argument for the sufficiency of the result based on the fact that the graph obtained by triangulating a simple polygon is three-colorable. Each triangle contains vertices of each color. A triangulation is a partitioning of the polygon into (non-overlapping) triangles. By placing guards at all vertices colored with the same color, it is ensured that the entire polygon is seen. The argument is completed by choosing that color which occurs least frequently, i.e. at most $\lfloor n/3 \rfloor$ times. The computational aspect of the problem was solved by Avis and Toussaint [1] who showed that such a placement for the guards could be determined in O(n log n) time.

Kahn, Klawe and Kleitman [12] studied the guard problem for polygons in which each edge is parallel to one of two given orthogonal axes (here taken to be the x-axis and y-axis): These polygons have been referred to as *rectilinear* or *orthogonal* and arise in many application areas such as image processing, VLSI-design and computer graphics. Kahn et al. showed that $\lfloor n/4 \rfloor$ guards are always sufficient (and for some polygons necessary) to guard any rectilinear polygon. The result is referred to as the *Rectilinear Art Gallery Theorem.*

The necessity of $\lfloor n/4 \rfloor$ guards for some rectilinear polygons is illustrated in Figure 1. There are on the order of $n/4$ "peaks", each peak being defined by 4 vertices and requiring one guard.

Figure 1 A polygon for which $\lfloor n/4 \rfloor$ guards are necessary.

The approach taken in [12] is to construct a four-colorable graph with the property that placing guards at all vertices of any one chosen color, will solve the guard placement problem. Then, by picking the least frequently occurring color, the $\lfloor n/4 \rfloor$ bound on the number of guards is obtained. The four-colorable graph is readily obtained once the given rectilinear polygon has been partitioned into convex quadrilaterals (see Figure 2 for an illustration). Such a partitioning joining only vertices of the polygon is referred to as a *convex quadrilaterization.* Notice that it is important for the quadrilaterals to be convex since, otherwise, it is not guaranteed that the guards placed arbitrarily inside a quadrilateral will be able to see the entire quadrilateral.

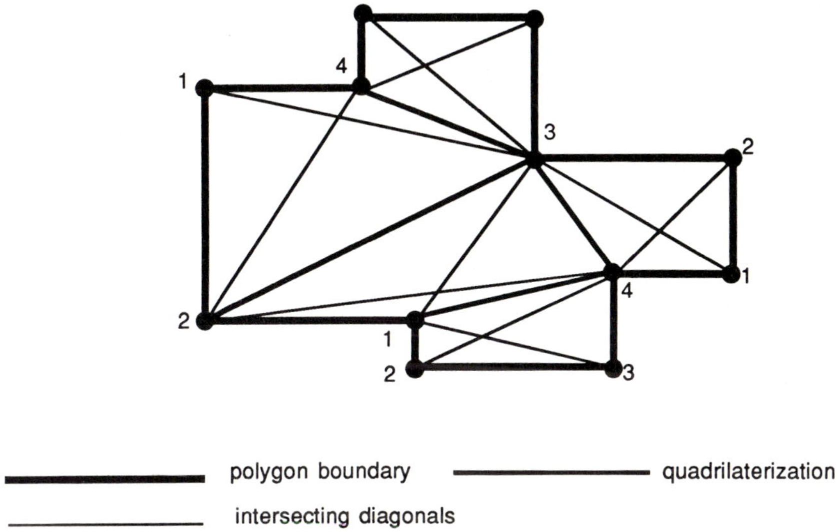

Figure 2 By inserting a pair of intersecting diagonals in each convex quadrilateral a four-colorable graph is obtained.

The hard part of their rather lengthy proof is to show the existence of such a convex quadrilaterization for any given rectilinear polygon. Quoting from their paper [12, p.197]: "In many places in our paper our proofs depend on certain configurations having particular properties which appear to follow obviously from the relevant definitions.... However, as is often true in geometric problems, despite their "obvious truth" it seems both time-consuming and tricky to provide rigorous proofs of their assertions."

The proof presented in [12] establishes the existence of a convex quadrilaterization for rectilinear polygons but does not lead immediately to an efficient algorithm for determining one. Here we give an alternate proof which provides also efficient algorithms for quadrilaterization. Since a four-coloring of a graph obtained from a convex quadrilaterization can be generated in linear time, an efficient solution is likewise obtained for the guard placement. The algorithms discussed in this paper have originally been presented in [21,22]. We will also show that by using the recent result by Tarjan and Van Wyck [25] our quadrilaterization algorithm can be further improved to run in O(n loglog$n$) time, for an arbitrary n-vertex, simple, rectilinear polygon. Subsequently, other algorithms (achieving the similar time bounds) have been developed for guard placement with or without quadrilaterization [11,17-20]. For these algorithms and for a thorough treatment of other art gallery theorems and algorithms the reader is referred to the recently published book by Joseph O'Rourke [20].

1.2 Two Natural Approaches

Whereas it is trivial to triangulate a simple polygon "by hand", it is much harder to manually find a convex quadrilaterization of a rectilinear polygon. We try to investigate why this is so. When solving the problem manually, even for rectilinear polygons with a small number of vertices, we often end up with concave quadrilaterals and/or triangles. In this case we must backtrack, possibly even to the first diagonal inserted. For example the natural heuristic to "cut off ears", as illustrated in Figure 3, may result in the creation of concave quadrilaterals.

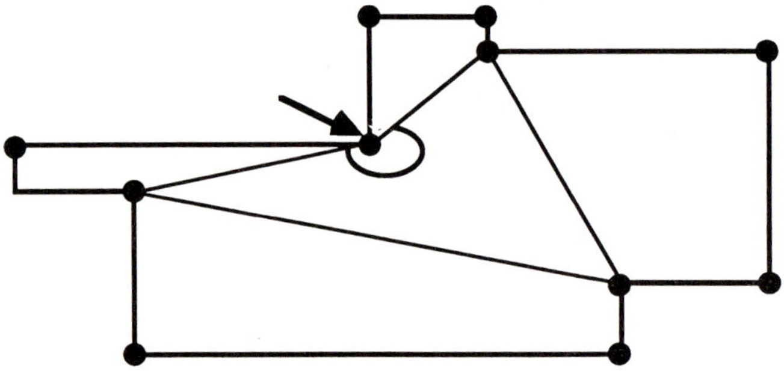

Figure 3 Illustrating the problem of creating possibly concave quadrilaterals.

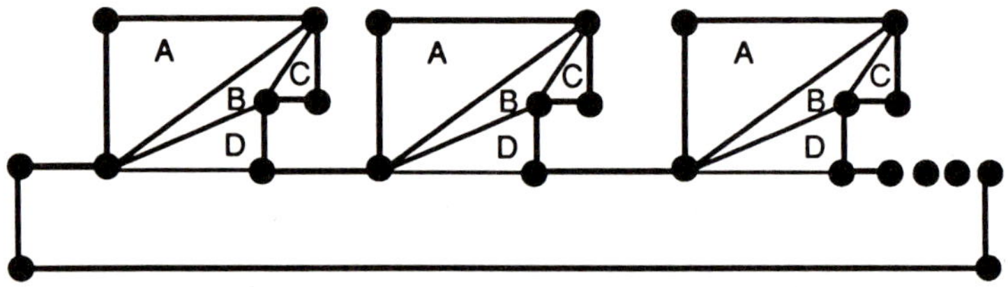

Figure 4 Creating a quadrilaterization by merging triangles from a given triangulation may result in as many as O(n) unmergeable triangles.

Another "natural approach" is to triangulate the polygon and simply erase every other diagonal thereby combining triangles to form convex quadrilaterals.

(No diagonal may be created or swapped). Since after completion of such a procedure triangles may remain which cannot be combined to form quadrilaterals, this approach may also be unsuccessful. As illustrated, in Figure 4, the number of such unmergeable triangles may be as large as O(n). In this Figure the entire rectilinear polygon P is composed of O(n) copies of a 6-vertex polygon whose triangles are labelled by A, B, C, D. The only two triangles that can be merged to form a convex quadrilateral are A and B. If in each such piece this diagonal is erased then triangle C remains unmergeable. Since this occurs repeatedly for every 6-vertex piece, as many as O(n) triangles will remain unmergeable.

Clearly, there is a "right" triangulation, since from a given quadrilaterization a triangulation is easily obtained by adding to each convex quadrilateral any of the two diagonals. However, finding such a triangulation (without first quadrilaterizing) is probably just as hard as finding the quadrilaterization directly; the problem remains open.

At this stage, the reader might be interested in knowing whether a quadrilaterization of a given rectilinear polygon unique. The answer to this question is: NO. In fact, some polygons even have an exponential number of different quadrilaterizations, as can be observed from the polygon illustrated in Figure 5.

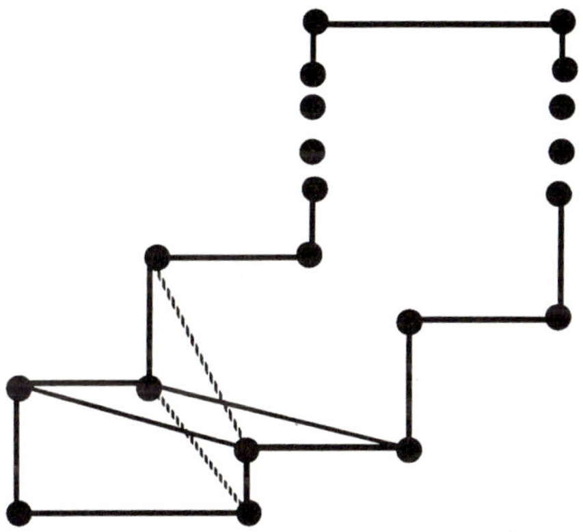

Figure 5 A rectilinear polygon with an exponential number of different quadrilaterizations. Each pair of consecutive reflex vertices on opposite chains has, as indicated, two choices independent on the choices for the reflex vertices of the remaining polygon.

2. OVERVIEW OF ALGORITHM

The algorithm described here parallels the development of triangulation algorithms for polygons. It has been shown that monotone polygons can be triangulated in linear time [10]. Since any simple n-vertex polygon can be partitioned into monotone pieces in time O(n log n) [10], an arbitrary simple polygon can be triangulated in O(n log n) time. We will recall the definitions of polygon, polygonal chain and monotonicity.

A *polygon* P is given by a list of its vertices $(p_1,\ldots,p_n)$ specified in clockwise order around the boundary of P and by its edges $(e_1,\ldots,e_n)$, where $e_i = p_i\, p_{i+1}$, $i=1,\ldots,n$. All operations are taken modulo such that $p_{n+1} = p_1$. We assume that all polygons are *simple*, i.e. no two non-adjacent edges are intersecting.

A *polygonal chain* $C_{i,j}$ is a sequence of consecutive vertices $p_i,\ldots,p_j$ of P. Such a chain $C_{i,j}$ is *monotone* with respect to a line l if the projections of the vertices p_k, for $k=i,\ldots,j$ on l and the vertices in $C_{i,j}$ are ordered in exactly the same way.

A polygon P is *monotone with respect to a line l* if the boundary of P can be partitioned into two chains $C_{i,j}$ and $C_{j,i}$ both of which are monotone in l. A polygon P is *monotone* if there exists a line l for which P is monotone.

Recently Tarjan et al. have reduced the O(n log n) bound for triangulation to O(n loglog n) [25]; leaving still open the question of whether there exists a linear-time algorithm for triangulation of a simple polygon.

Here we will use a decomposition-based approach for solving the quadrilaterization problem (which itself is a particular decomposition problem). For a survey on decomposition techniques in polygons see [13, 20]. We will first show how monotone rectilinear polygons can be quadrilaterized. Unlike for triangulation, any algorithm for partitioning a given rectilinear polygon into monotone rectilinear polygons may create Steiner points. These Steiner points may become endpoints of diagonals if an arbitrary quadrilaterization algorithm for monotone rectilinear polygons is applied to the monotone pieces. As illustrated in Figure 6 it may happen that almost all diagonals have such Steiner points as endpoints.
Two approaches can thus be taken:

(1) Partitioning into Monotone Non-rectilinear Pieces
Not every monotone polygon can be quadrilaterized. A possible approach is to define a class of monotone polygons which is quadrilaterizable and then show how to partition the given rectilinear polygon into such monotone pieces. Such a class of polygons has been described in [24] and was rediscovered in [17]. For our discussion here we take a different approach.

(2) Covering with Monotone Rectilinear Pieces
We will show that a given rectilinear polygon can be covered with monotone rectilinear pieces in such a way that a subsequent quadrilaterization of the monotone rectilinear pieces will give a quadrilaterization of the polygon. This will be true only if the algorithm for quadrilaterization of monotone rectilinear polygons developed here is used, since other algorithms do not necessarily have the particular properties required for the covering approach.

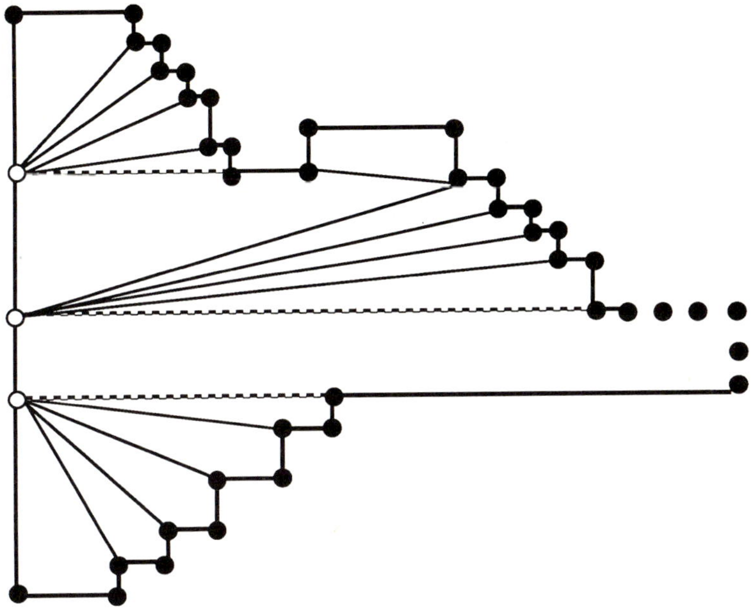

Figure 6 A partitioning of a rectilinear polygon into monotone rectilinear polygons creating Steiner points. Subsequent quadrilaterization of these may create many diagonals whose endpoints are Steiner points.

3. QUADRILATERIZING PYRAMIDS

An essential "building block" for rectilinear polygons is the *pyramid*, which is a rectilinear polygon monotone in both the *x*-direction and the *y*-direction and having exactly one bottom edge. The two *y*-monotone chains of a pyramid are called the *left* and *right staircase,* respectively. Figure 7 illustrates a pyramid and its quadrilaterization obtained by Algorithm *Pyramid Quadrilaterization* described below.

Pyramids are quadrilaterized in the following way: Create two lists, called *L* and *R*, containing vertices for the left and right reflex vertices, respectively, sorted in order of descending *y*-coordinates. The lists are scanned in parallel and each reflex vertex of one chain is joint to the next lower reflex vertex on the other chain. Those reflex vertices for which no lower reflex vertex on the other chain exists are joint to the vertex with minimum *y*-coordinate on the other chain.

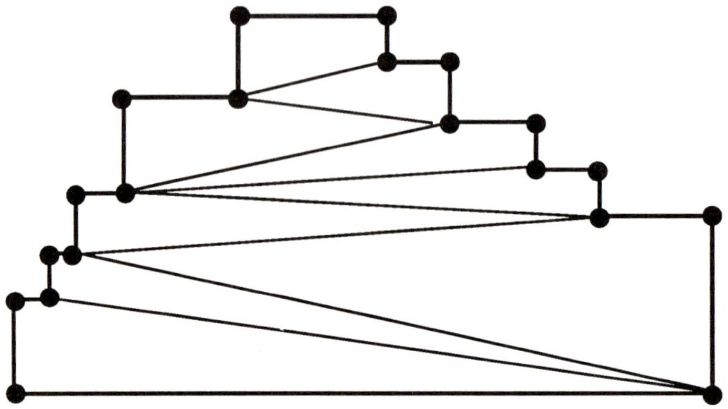

Figure 7 A quadrilaterization of a pyramid.

Algorithm ***Pyramid Quadrilaterization***

Input: A pyramid P.
Output: A convex quadrilaterization of P.

Step 1: Starting with the y_{max}-edge construct lists L and R:
$L := l_1, \ldots, l_n$ of all reflex vertices on the left staircase in descending order of y-coordinates;
$R := r_1, \ldots, r_m$ of all reflex vertices on the right staircase in descending order of y-coordinates.

Step 2: **repeat**
in each list
join each vertex to the next lower vertex in the other list
until one list is empty

Step 3: join all remaining vertices to the vertex with minimum y-coordinate located on the other staircase.

Remark 3.1 If the y-coordinates of the vertices $l_i, l_{i+1}, \ldots, l_k$ are between those of r_j and r_{j+1} then the above algorithm joins all vertices $l_i, l_{i+1}, \ldots, l_k$ to r_{j+1}.

Lemma 3.2 Algorithm *Pyramid Quadrilaterization* quadrilaterizes pyramids in linear time.
Proof: Let P be an n-vertex pyramid. In Step 1 of the algorithm two lists of reflex vertices sorted by y-coordinates are created. If both lists are empty, the polygon is already a convex quadrilateral and the algorithm stops. Let therefore at least one of the lists be non-empty. The algorithm creates polygons, say $Q_1, \ldots, Q_d$. We show that:

(a) $Q_1, \ldots, Q_d$ are quadrilaterals.
(b) $Q_1, \ldots, Q_d$ are convex;
(c) The polygons $Q_1, \ldots, Q_d$ are pairwise non-intersecting (properly);

(d) The union of all $Q_1,...,Q_d$ is the entire pyramid P;
(e) The algorithm runs in linear time.

(a) From the algorithm it is clear that every pair of consecutive reflex vertices on the left (right) staircase is joined to a common vertex on the right (left) staircase, therefore the components are quadrilaterals.

(b) Let v be reflex vertex, say on the left staircase and refer to Figure 8. The concavity at v is broken by any diagonal located in region B. The algorithm inserts a diagonal vw such that
(i) w is located on the other staircase, therefore vw is not located in region A.
(ii) v has a higher y-coordinate than w; therefore vw is not located in region C.
All other reflex vertices are handled in a similar way. Since all other angles in the quadrilaterals are either $\pi/2$ or less, the quadrilaterals are convex.

(c) Since the vertices are scanned in order by decreasing y-coordinates, it follows trivially that no two quadrilaterals intersect.

(d) Each reflex vertex is contained in at least two quadrilaterals. Each convex vertex (except r, s (p,q) the top and (bottom) vertices) is located between two reflex vertices and is therefore contained in the respective quadrilateral. The vertices r,s, (p,q) are included in the highest (lowest) quadrilateral. Therefore the union over all quadrilaterals is equal to the entire polygon.

(e) The running-time of Step 1 is proportional to the sum of the elements in both lists and is therefore O(n). Since pyramids are monotone in the y-direction no sorting is needed. Adding an edge is Steps 2 and 3 takes constant time. Therefore the total run-time is O(n). q.e.d.

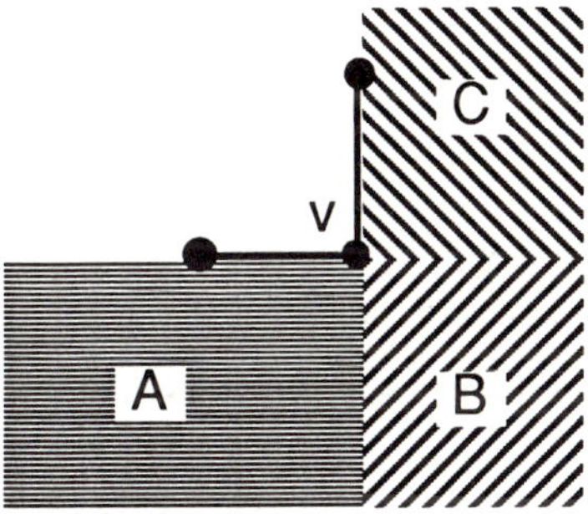

Figure 8 Indicating locations for points to which a reflex vertex may be connected.

Eroded Pyramids

Algorithm *Pyramid Quadrilaterization* will be utilized in the quadrilaterization algorithm for monotone rectilinear polygons described below. We will, however, apply Algorithm *Pyramid Quadrilaterization* to a slightly more general class of polygons than pyramids, referred to as *eroded pyramids.*

Let $p=(x_p,y_q)$ and $q=(x_q,y_q)$ be two points in the plane. The point $p\#q$ is defined as the point (x_p, y_q), i.e. the point with x-coordinate equal to that of p and y-coordinate equal to that of q. Analogously $q\#p$ is defined. (See Figure 9)

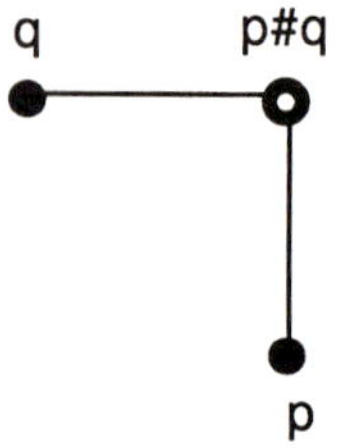

Figure 9 The Steiner point *p#q* is the point with *x*-coordinate equal to that of *p* and *y*-coordinate equal to that of *q*.

A horizontal edge *e* of a rectilinear polygon *P* is called a *horizontal top (bottom)-edge* if the interior of *P* is below (above) *e* .
We generalize this notion: Let *p, q* be the endpoints of an edge *e* of a simple y-monotone polygon *P* and let *p* be located below *q*. Then *e* is called a *top edge* if the point *p#q* is outside *P* (see Figure 10). By replacing *p* by *p#q*, the vertical edge containing *p* is elongated and one horizontal top edge connecting *q* to *p#q* is created. To avoid unnecessary case analyses, horizontal top edges will sometimes be considered as special cases of top edges. The term *bottom edge* is analogously defined.

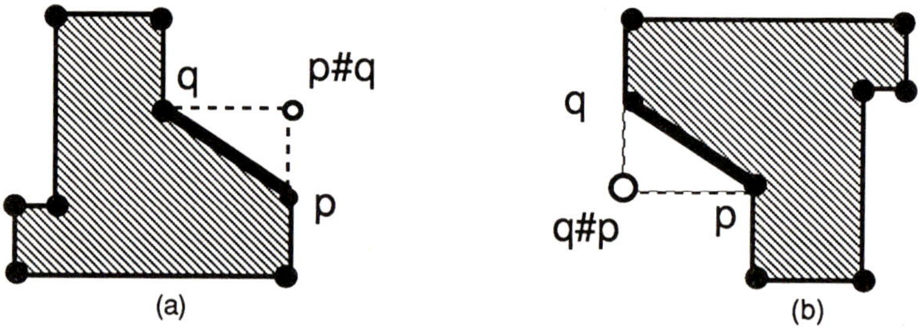

Figure 10 In (a) the edge *qp* is a top edge, while *pq* is a bottom edge in (b).

Let *P* be a pyramid with *rs* and *pq* as top and bottom edges, respectively. A polygon *P* is called *rectilinearly convex* if each horizontal or vertical line that intersects *P*, does so in at most one connected component. A rectilinearly convex polygon P_W is called an eroded *pyramid* if all its even-numbered edges are vertical and all its odd-numbered edges are top edges possibly non-horizontal; the odd-numbered y_{min}-edges (y_{max}-edges) are allowed to be non-horizontal only, if both endpoints remain below (above) the lowest (highest) reflex vertex in *P*.

It follows that eroded pyramids are rectilinearly convex, *edge-visible* polygons with an alternating sequence of vertical edges and top edges. Recall that a polygon *P* is edge-visible if it contains an edge *e* such that for every point *x* in *P* there exists a point *y* in *e* with the property that the line segment *xy* lies in *P*. From the definition of a top edge it follows that any eroded pyramid is embedded in a (rectilinear) pyramid. See Figure 11 for an illustration. It is easy to see that with Algorithm *Pyramid Quadrilaterization* these eroded pyramids can also be decomposed into convex quadrilaterals since the interior angles of the rectilinear embedding which are $\pi/2$ are still less than π in the eroded pyramid and are therefore convex.

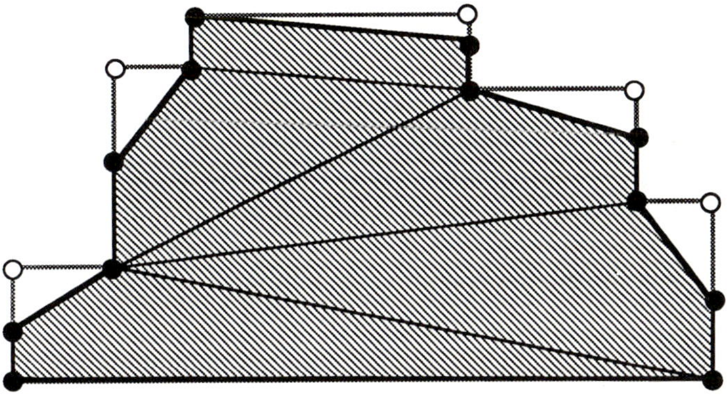

Figure 11 An eroded pyramid and its quadrilaterization; the original pyramid before erosion.

4. QUADRILATERIZING MONOTONE POLYGONS

In this section we present a linear-time quadrilaterization algorithm for monotone rectilinear polygons. In [24] it has been shown that any monotone rectilinear polygon is monotone to the y-axis and/or the x-axis and a given rectilinear polygon can be efficiently tested for monotonicity, in linear time. We can therefore assume that the given rectilinear polygon is monotone with respect to the y-direction. The algorithm presented here results from merging algorithms designed by Hossam ElGindy [7] and by the first author [22].

The algorithm scans the horizontal edges of the given polygon from top to bottom. Whenever a top edge is encountered the edge is pushed onto a stack which will contain the non-vertical edges of an eroded pyramid. A bottom edge may trigger one or two diagonals to be output; furthermore, if the bottom edge joins the endpoints of the chains defining the eroded pyramid contained in the stack, the pyramid quadrilaterization algorithm is evoked and the stack is emptied. Whether one or two edges are printed depends on whether the current bottom edge and the lowest edge in the stack touch the same chain or are on opposite chains, respectively. (Edges may touch the left chain, right chain or both chains.) We are now able to state the algorithm whose pseudocode presentation has been improved thanks to Joseph O'Rourke.

Algorithm Quadrilaterization of Monotone Rectilinear Polygons

Input : A rectilinear polygon P monotone in the y-direction
Output : The diagonals defining the quadrilaterization of P

sort the horizontal edges of P in decreasing y-coordinates;
let $(e_1,...,e_n)$ be the sorted edge-list;
Push e_1 onto the stack {*initialize the stack*}
$i := 2$;

```
while i<n do
begin
case e_i
(A): top edge (in the remaining polygon)
  begin
      Push e_i onto the stack;
      increment i
  end
(B): bottom edge (in the remaining polygon)
  begin
      let e := Pop(Stack);
      if at least one of the endpoints of e is on the same chain as that of e_i
      then
      (B1)
      begin
         join those endpoints of e and e_i which are not connected by a
         diagonal called d;  {a convex quadrilateral is cut off}

         if d  is a top edge in the remaining polygon  then
         begin
             Push d  onto the stack;
             increment i
         end
         else
             e_i := d; {the loop will be executed with d replacing e_i  and without
                   advancing i }
      end

      (B2)  {no endpoint of e is on the same chain as e_i}
      else
      begin
         join e and e_i  by two diagonals to form a convex quadrilateral;
         Push the higher diagonal onto the stack;
         Quadrilaterize eroded pyramid in stack;
         increment i;
         Push lower diagonal ontoi the stack;
      end
  end
end
(Final  Step)
Push e_n onto the stack
Quadrilaterize eroded pyramid in stack
```

Before establishing the correctness of the algorithm it is convenient to introduce the notion of shadow. The *shadow* of an edge is the set of all points interior to *P* that are horizontally visible from some (not necessarily the same) point on the edge. More formally, the shadow of an edge *e* of polygon P, denoted by *SH(e, P)* is the set of all points *s* in P such there exists a point *z* on *e* with the same coordinate as *s* and such that the line segment *xz* lies in P. Thus the shadow of a horizontal edge is a line segment.

Theorem 4.1 Algorithm *Monotone Quadrilaterization* quadrilaterizes monotone rectilinear polygons in linear time.
Proof It is easy to observe that the amount of work per edge as done by the algorithm is constant and thus the time complexity is linear.
To establish the algorithm's correctness we will show that the stack, maintained by the algorithm, satisfies the following three properties, any time the algorithm starts to execute the body of the while-loop. This will hold regardless of whether the edge e_i is an edge of the input polygon or an edge of the polygon created during the execution of the algorithm by cutting off convex quadrilaterals.

__Stack Property__:

(1) The stack contains all top edges of an eroded pyramid and only those edges.
(2) The edges contained in the stack have non-intersecting shadows.
(3) The edge e_i to be processed and the eroded pyramid contained in the stack are connected by a vertical edge of *P*.

In particular this will imply that when the edge with minimum *y*-value is encountered, the stack content together with this edge define an eroded pyramid. The final call to Procedure *QuadrilaterizePyramid* will then quadrilaterize this pyramid. The proof is by induction on the number of edges in *P*. The stack property clearly holds after the initialization phase of the algorithm. Assume therefore that the Stack Property holds when edge e_i, $i > 2$ is to be processed.

Case (A) : e_i is a top edge (of the remaining polygon)
By induction, (3) holds and thus edge e_i is vertically connected to the eroded pyramid contained in the stack. The eroded pyramid can thus be augmented by pushing edge e_i onto the stack and (1) remains true. Furthermore, since horizontal and vertical edges of rectilinear polygons alternate, Property (3) holds. Finally, the edges of *P* are scanned in order sorted by *y*-value and therefore no shadow of an edge contained in the stack can have the same *y*-value as e_i implying that (2) remains true.

Case (B): e_i is a bottom edge (of the remaining polygon)
Denote by *e* the top edge whose endpoint has lowest *y*-value among all edges currently residing in the stack. Edge *e* is popped off the stack and depending on its location further processing is done.

Case (B1): At least one of the endpoints of e is on the same chain as e_i
By Property (3), the edges *e* and e_i are connected by a vertical edge. The algorithm inserts the diagonal connecting the as yet unconnected endpoints of *e* and e_i; this creates a convex quadrilateral. The rectilinearity of the polygon ensures that Property (3) remains true.

Subcase (I): the endpoints of e are on opposite chains
Since no polygonal edges except the extremal edges e_1 and e_n connect both chains, *e* either results from Step (B_2) or *e* is e_1. An inspection of the algorithm reveals that the Push operation for *e* must have been preceded by a call to the *Pyramid Quadrilaterization* which as a side-effect empties the stack. Since also edge *e* has been popped, in both cases, the stack will be empty. The diagonal

connecting e and e_i is a top edge for the remaining polygon and is pushed onto the stack and a situation similar to the initialization step is obtained. The stack properties thus hold. The situation is illustrated in Figure 12.

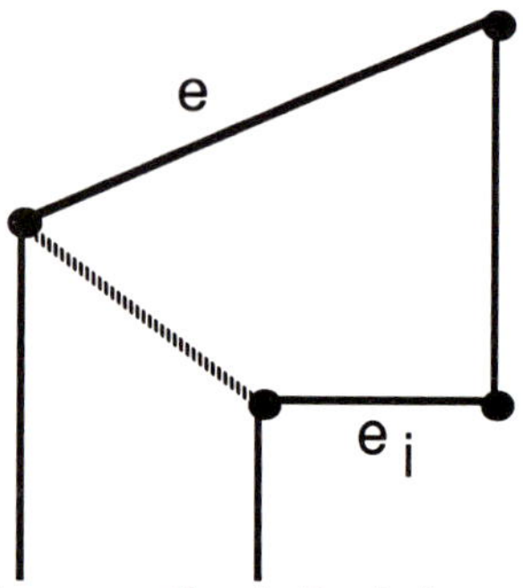

Figure 12 Diagonal e is connecting both chains; e and edge e_i form a convex quadrilateral.

Subcase (II): both edges e and e_i are on the same chain

For an illustration of this case see Figure 13. By induction, Property (3) holds and thus the algorithm can correctly insert the diagonal joining e and e_i to form a convex quadrilateral. In case the new diagonal is a top edge in the remaining polygon, it is pushed onto the stack. By induction Property (3) then ensures that this augmentation does not violate Property (1). In case that the diagonal is a bottom edge in the remaining polygon, no push operation is executed and thus clearly (1) must hold. It remains to show that in both cases (2) and (3) remain true.

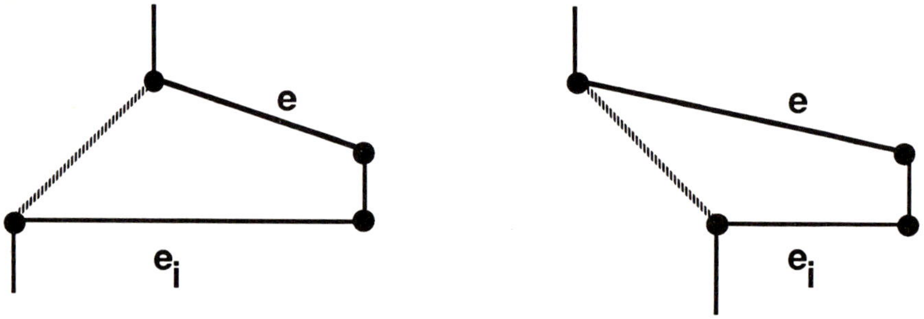

Figure 13 The edges e and e_i are on the same chain.

By induction, Property (2) holds prior to the examination of edge e_i. Since by the choice of e, no top edge on the stack has a vertical coordinate between that of e and e_i, Property (2) remains satisfied even if the diagonal, d joining e and e_i is inserted. It is easy to see that Property (3) holds in case d is a top edge and pushed onto the stack. Otherwise, d is a bottom edge, and the algorithm neither pushes d

onto the stack, nor does it advance to the next edge. It replaces e_i by d for the next execution of the loop body. If, for this iteration, d and the new pyramid edge popped off the stack are on opposite chains the argument is as in case (B2). Otherwise, the validity of this is easily observed by "rectilinearization" of d as indicated in Figure 14.

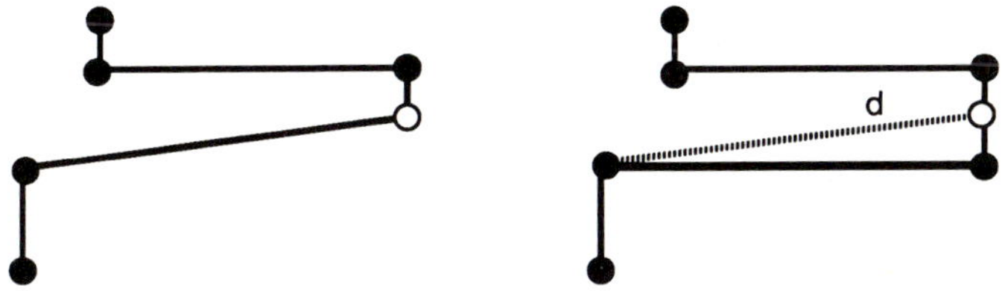

Figure 14 Rectilinearization of *d*.

Case (B2): Edge e and e_i are on opposite chains
The discussion is similar whether e_i is horizontal or not. W.l.o.g. let $e=cd$ be on the left chain and $e_i=ab$ be a non-horizontal bottom edge on the right chain. By the choice of e and Stack Property (2) the region interior to P and bounded by $R := (a,b,c\#b,c,d,a\#d)$ is empty (see Figure 15).

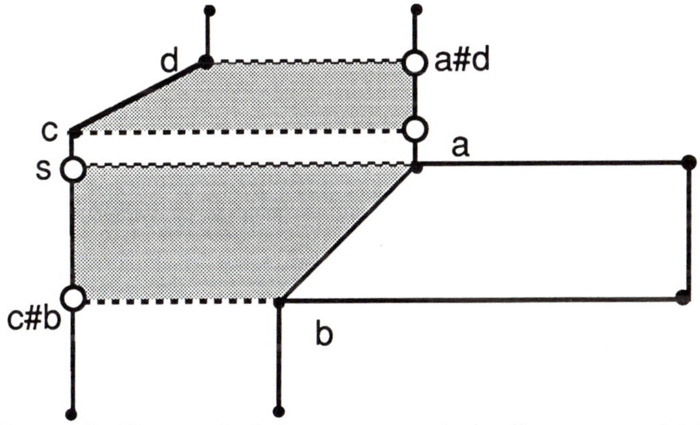

Figure 15 The two shaded areas are empty, i.e. they are completely in the interior of the polygon.

The algorithm performs three steps
(i) Push diagonal *ad*
(ii) Quadrilaterize pyramid in stack
(iii) Push diagonal *bc*

After Steps (i) and (ii) the stack is emptied and (iii) initializes a new eroded pyramid. Since by the above region R is empty the point pairs a,d and b,c are visible. We will now argue that ad is a diagonal of the quadrilaterization of P. Let *Pyr* denote the pyramid contained in the stack prior to the Pop-operation of the pyramid-edgee completed by an artificial horizontal bottom edge connecting *a* to the Steiner-point *s:=c#a*. A quadrilaterization of this pyramid will insert the diagonal *ad* (see Remark 3.1). This justifies (i). All other diagonals inserted for this quadrilaterization are identical to that produced by (ii) and thus (ii) is correct. The quadrilateral *d, a, s, c* will remain convex even if *sc* and *as* are replaced by *ab* and *bc*. Thus the entire region up to *bc* is correctly quadrilaterized. By the Push-operation of *bc* in (iii), a new eroded pyramid is started for which the three Stack Properties hold. q.e.d.

Properties of Algorithm Monotone Decomposition

The following two properties of the above algorithm are crucial for the Algorithm *QuadrilaterizeRectilinearPolygon* presented below. They follow directly from the description of the algorithm.

Property 4.2: Let *ab* be at the top of the stack when bottom edge *cd* is encountered. If *ab* and *cd* are on the same chain then Algorithm *MonotoneQuadrilaterization* inserts one diagonal such that *a,b,c,d* forms a convex quadrilateral.

Property 4.3: Let *ab* and *cd* be two edges consecutive on one of the monotone chains of the polygon. The reflex vertices of *ab* and *cd* are joined to a common vertex via diagonals inserted by Algorithm *MonotoneQuadrilaterization.*

5. QUADRILATERIZING RECTILINEAR POLYGONS

A rectilinear polygon *P* is *y*-monotone if and only if it contains no *horizontal peaks* where a horizontal peak is a horizontal edge whose endpoints are both reflex. A horizontal peak is a *bottom peak* if it is a bottom edge; the term *top peak* is analogously defined.

Our approach is to eliminate horizontal peaks such that the Steiner points created during that process will never occur as endpoints of diagonals inserted by the subsequent calls to Algorithm *MonotoneQuadrilaterization.*

We first define visibility between edges. Edge *ab* is said to be (partially) visible from edge *cd* if there exists a point *w* on *ab* and a point *z* on *cd* such that w and z are visible. For a discussion of variants of edge-visibility see e.g. [3]. We will call a triangle (or quadrilateral) formed by boundary points of a polygon *P* *empty* if it located entirely inside *P.*

Lemma 5.1 Let *ab* be a bottom peak in a rectilinear polygon *P* and *cd* the lowest horizontal edge located above *ab* and visible from *ab.*

(a) If *cd* is a bottom edge then either triangle Δ *a,b,c* or Δ *a,b,d* is empty.

(b) If *cd* is a top edge then the convex quadrilateral $\Diamond$ *a,b,c,d* is empty.

Proof (a) If *cd* is a bottom edge located above *ab* then clearly at most one of its endpoints can see *ab*. The remainder then follows from the fact that: If a point *x* in a polygon *P* is visible from the endpoints of an edge then it is visible from any point on the edge [2].

(b) Now let *cd* be a top edge. We show that the entire quadrilateral ◊ *a,b,c,d* is empty. By assumption *cd* is visible from *ab*, i.e. some point *q* on *cd* is visible from some *p* on *ab*. Consider the triangle Δ *q,a,b*. This triangle is empty, since, otherwise, there must exist horizontal edges of *P* which intersect *aq* or *bq* and the lowest such edge is visible from *ab* contradicting the choice of *cd*. By a similar argument it can be shown that any point on *cd* is visible from *ab* and thus the entire quadrilateral ◊ *a,b,c,d* is empty. q.e.d.

Let *ab* be a bottom peak and *cd* the lowest horizontal edge (above *ab*) that is visible from *ab*. The *rectangles defined* by *ab*, *cd* and the vertical edges facing *ab* and *cd* are illustrated in Figure 16.

Corollary 5.2 Under the assumptions of Lemma 5.1 the rectangles defined by *ab* and *cd* are empty.

We will first state the rule for decomposing a rectilinear polygon *P* into two rectilinear polygons with one less horizontal peak. We give the rule referring to Figure 16(a), (b), (c), the other cases are handled analogously.

<u>**Decomposition Rule:**</u>

Case (a), (b) P_{right} := (*a, w, c, d, ...*) and P_{left} := (*c, u, v, b, ...*)

Case (c) P_{right} := (*a, w, x, d, ...*) and P_{left} := (*c, u, v, b, ...*)

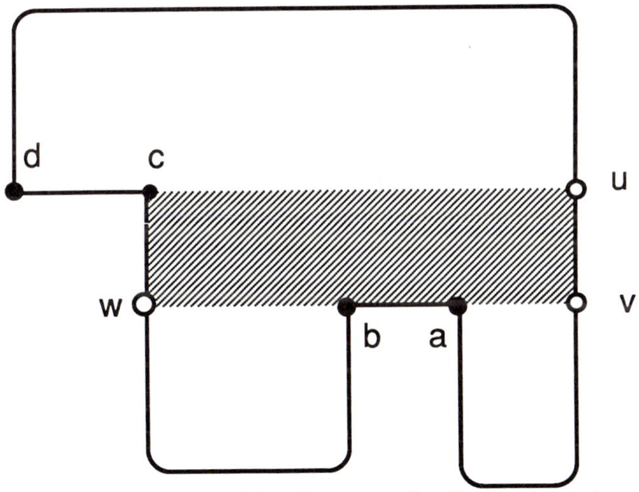

Case (a): *ab* is a bottom peak *cd* is a bottom edge.

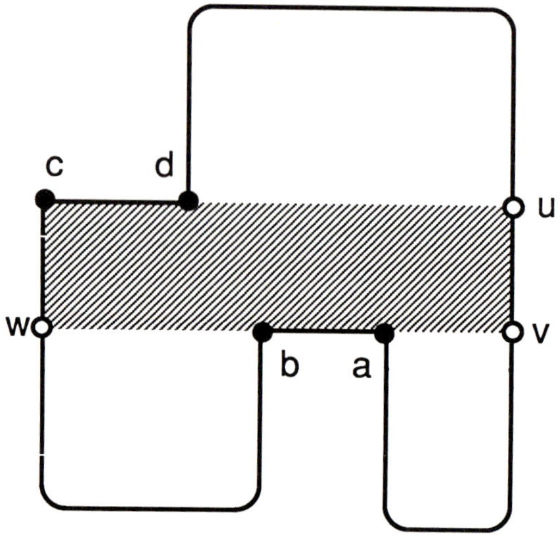

Case (b): *ab* is a bottom peak *cd* is a top edge.

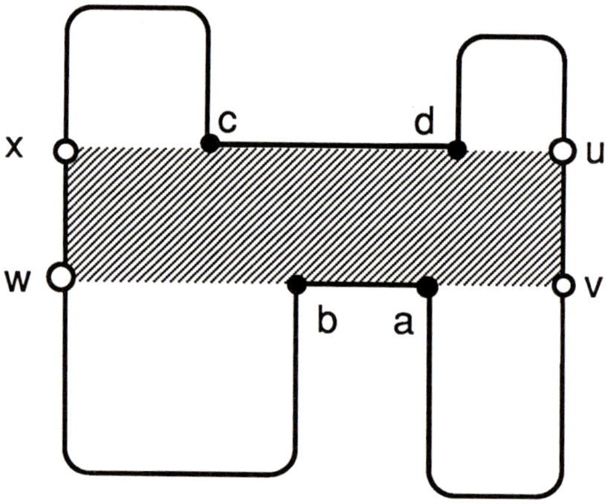

Case (c): *ab* is a bottom peak *cd* is a top peak.

Figure 16 The rectangles defined by the edges *ab* and *cd* in the interior of *P* are empty.

Algorithm *CoverRectilinearPolygon*
Input: A rectilinear polygon $P = (p_1, \ldots, p_n)$
Output: A covering of P with monotone rectilinear polygons

for each horizontal bottom (top)-peak **do**
find next higher (lower) visible edge
eliminate peak using Decomposition Rule

Lemma 5.3 (i) Algorithm *CoverRectilinearPolygon* covers a given rectilinear polygon with monotone rectilinear polygons.
(ii) If these rectilinear polygons are quadrilaterized using Algorithm *MonotoneQuadrilaterization* then a quadrilaterization of the given rectilinear polygon is obtained.

Proof (i) Clearly, Algorithm *CoverRectilinearPolygon* creates a covering of the given rectilinear polygon. Furthermore, since each of the polygons so created is, by construction, rectilinear and contains no horizontal peaks, a covering with monotone rectilinear polygon is obtained.

The more interesting part is to show (ii), i.e. to show that if each monotone rectilinear polygon is quadrilaterized separately using Algorithm *Monotone Quadrilaterization* a quadrilaterization of the original polygon is obtained. We must show that

(1) no pair of diagonals inserted by a quadrilaterization of different monotone pieces intersects.

(2) No diagonal contains a Steiner point as one of its endpoints.

We analyze each of the three cases illustrated in Figure 16 above. The cases that the next higher visible edge is on the right chain or that the horizontal peak is a top peak can be treated analogously. The proof is by induction on the number of peaks; we give the induction step only.

Case(a) By the decomposition rule the two resulting polygons are:
$P_{right} := (a, w, c, d, \ldots)$ and $P_{left} := (c, u, v, b, \ldots)$

In P_{left}, edge *cu* is a top edge and *vb* is a bottom edge. By Lemma 5.1 no horizontal edge in P_{left} has *y*-coordinate between that of *cu* and *vb*. Thus when Algorithm *MonotoneQuadrilaterization* examines the bottom edge *vb* the edge on the Top_Of_Stack is edge *cu*. By Property 4.2, the diagonal *cb* will thus be inserted creating the convex quadrilateral *c,u,v,b*. This quadrilateral does not contain any other diagonal inserted during the quadrilaterization of P_{left}. After P_{left} has been quadrilaterized completely, we can cut this quadrilateral off and the remaining quadrilaterals form a convex quadrilaterization of *b, ..., c*.

Now we turn to the quadrilaterization of P_{right}, as performed by Algorithm *MonotoneQuadrilaterization*. In P_{right} the edges *cd* and *aw* are consecutive on the same chain. Thus, by Property 4.3, their reflex vertices *c* and *a* are joined to the same vertex, *z*, to form the convex quadrilateral *c,z,a,w*. By cutting off the triangle *b,w,c* from this quadrilateral, a new quadrilateral *z,a,b,c* is built; this new quadrilateral remains convex. We thus obtain a convex quadrilaterization of the polygon *b,c,...,a*.

By combining the above two quadrilaterizations, a quadrilaterization of *P* is obtained. Notice that the cutting off operations are purely of a conceptual nature and that all diagonals of the quadrilaterizations of P_{left} and P_{right} are diagonals for the quadrilaterization of the original polygon *P*.

Case (c) The decomposition rule for this case is: P_{right} := (*a, w, x, d, ...*) and P_{left} := (*c, u, v, b, ...*). By Lemma 5.1 the quadrilateral defined by *cd* and *ab* is empty. Thus since *cd* and *ab* are portions of edges in both P_{right} and P_{left} , they will be consecutive in the order in which the edges are processed by the quadrilaterization algorithms for P_{left} and P_{right}. Algorithm *MonotoneQuadrilaterization* will produce the quadrilateral *c,u,v,b* with diagonal *cb* in P_{left} and the quadrilateral *a,w,x,d* with diagonal *ad* in P_{right}. When cutting off these quadrilaterals the resulting quadrilaterizations are disjoint and cover *P* except for the quadrilateral *c,d,b,a* which is convex since *cd* and *ab* are parallel. As above, a quadrilaterization for the original polygon is obtained from the quadrilaterizations of the two monotone pieces which contains all diagonals inserted by the two calls to Algorithm *MonotoneQuadrilaterization*. This case is similar in spirit to the notion of reduction using "neighboring edges" as discussed in [12].

Case (b) Similar.
q.e.d.

It remains to discuss the time required for Algorithm *CoverRectilinearPolygon*, i.e. how fast can we find for each bottom(top)-peak in the polygon, the next higher (lower) edge and visible from the peak. Several algorithms can be utilized. We will sketch four of them and refer the reader to the respective literature.

(1) **"Brute Force"**
Find the lowest horizontal edge that is located above the current peak and is visible from the peak. This step may require $O(n^2)$, in worst case, even if the horizontal edges are sorted by y-coordinates.
Total Time: $O(n^3)$.

(2) **"Straight Forward"**
Compute the visibility polygon from each of the endpoints of the given peak. Then choose among all edges located completely or partially in the visibility polygon the one with minimum distance in *y*-coordinate above (or below) the peak. Since the visibility polygon from a point can be determined in linear time [5,6,15,22] the total time required for the algorithm is $O(n^2)$.
Total Time: $O(n^2)$.

(3) **"Regularization"**
Lee and Preparata [14] have developed a technique for *regularizing* simple polygons. Regularization, which requires $O(n \log n)$ time for a simple *n*-vertex polygon, consists of adding diagonals in such a way that each vertex (except the two vertices extremal in the *y*-direction) is connected to at least one vertex below it and to one vertex above it. In order to be able to insert

diagonals the algorithm must maintain visibility information inside the polygon. This visibility information is maintained in a balanced tree, while the algorithm scans the vertices of the polygon in sorted order by *y*-coordinates (plane-sweep approach). How regularization can be used for the covering step is described in detail in [23].
Total Time: O(*n* log *n*).

(4) **"Trapezoidization"**
A *trapezoidization* of a polygon is a partitioning into trapezoids, where all trapezoids have their parallel edges also parallel to each other. (Trapezoids may degenerate to triangles.) The trapezoidization problem is to find a minimal set of disjoint trapezoids covering the given polygon. Trapezoidization has been discussed by Fournier and Montuno [9], where it is shown that given a trapezoidization, a triangulation can be obtained in linear time. It has been shown by Tarjan and VanWyck that a trapezoidization of a simple *n*-vertex polygon can be obtained in O(*n* loglog *n*) time [25]. Given a trapezoidization a linear-time scan over the trapezoids suffices to determine all peak-edge pairs necessary to finding the covering of the given rectilinear polygon into monotone rectilinear polygons.
Total Time: O(*n* log log *n*).

We thus obtain:

Theorem 5.4 A given *n*-vertex rectilinear polygon can be covered by monotone rectilinear polygons in O(*n* loglog *n*) time.
Proof: Follows directly from the result by Tarjan and Van Wyck [25] and the fact that the edge pair necessary for the covering step can be obtained from the trapezoidization of the given rectilinear polygon. q.e.d.

We have shown that any given rectilinear polygon can be covered by monotone rectilinear polygons. The covering was chosen such that when the Algorithm *Monotone Quadrilaterization* is applied to each piece no diagonal is created that joins to a Steiner point created in the covering step. Thus this quadrilaterization of the monotone pieces is a quadrilaterization of the entire rectilinear polygon.

Theorem 5.5 A given *n*-vertex rectilinear polygon can be quadrilaterized in O(*n* loglog *n*) time.
Proof: Follows from Lemma 5.3 and Theorem 4.1. q.e.d.

Corollary 5.6 At most $\lfloor n/4 \rfloor$ guards can be placed in an arbitrary, *n*-vertex, rectilinear polygon in O(*n* loglog *n*) time.

Figure 17 illustrates a complete quadrilaterization of a simple rectilinear polygon and a possible guard placement.

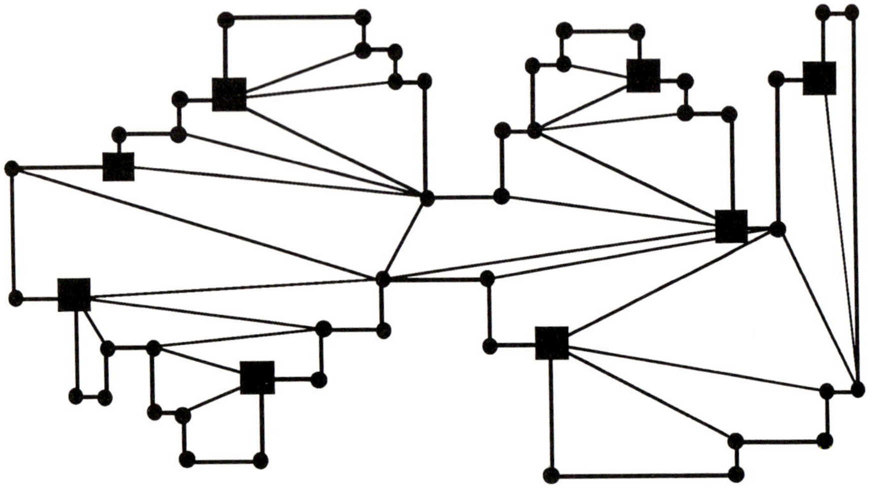

Figure 17 A quadrilaterization produced by the algorithm and a possible guard placement at vertices indicated.

OPEN PROBLEMS

Lee and Lin [16] have shown that the problem of placing the minimal number of guards needed to guard a given simple polygon (without holes) is NP-hard. The problem remains open in the context of rectilinear polygons.

ACKNOWLEDGEMENTS

The first author is indebted to Joseph O'Rourke for helpful discussions.

REFERENCES

[1] Avis, D. and G.T. Toussaint, An Efficient Algorithm for Decomposing a Polygon into Star-Shaped Polygons, *Pattern Recognition* , Vol. 13, 295-298 (1981).
[2] Avis, D. and G.T. Toussaint, An Optimal Algorithm for Determining the Visibility of a Polygon from an Edge, *IEEE Trans. on Computers,* Vol. C-30, No.12, 910-914 (1981).
[3] Avis, D., T. Gum and G.T. Toussaint, Visibility between two Edges of a simple Polygons, *The Visual Computer*, Vol. 2, 342-257 (1986).
[4] Chvátal, V., A Combinatorial Theorem in Plane Geometry, *Journal of Combinatorial Theory, Ser. B*, Vol. 18, No. 1, 39-41 (1975).
[5] Dean, J.A. and J.-R. Sack, Efficient Hidden-Line Elimination by Capturing Winding Information, *Proceedings of the 23rd Annual Conference on Communication, Control, and Computing,* 496-505 (1985).
[6] ElGindy, H. and D. Avis, A Linear-Time Algorithm for Computing the Visibility Polygon from a Point, *Journal of Algorithms,* Vol. 2, No. 2, 186-197 (1983).
[7] ElGindy, H., Personal communication.
[8] Fisk, S., A Short Proof of Chvátal's Watchman Theorem, *Journal of Combinatorial Theory, Ser. B,* Vol. 24, No.3, 374 (1978).

[9] Fournier, A. and D.Y. Montuno, Triangulating Simple Polygons and Equivalent Problems, *ACM Transactions on Graphics,* Vol. 3, No. 2, 153-175 (1984).
[10] Garey, M.R., D.S. Johnson, F.P. Preparata, and R.E. Tarjan, Triangulating a Simple Polygon, *Information Processing Letters,* Vol. 7, No. 4 175-179 (1978).
[11] Györi, E., A Short Proof of the Rectilinear Art Gallery Theorem, *SIAM Journal of Algebraic and Discrete Methods,,* Vol. 7, 452-454 (1986).
[12] Kahn, J., M. Klawe, and D. Kleitman, Traditional Galleries Require Fewer Watchmen, *SIAM Journal of Algebraic and Discrete Methods,* Vol. 4, No.2, 194-206 (1983).
[13] Keil, M.J. and J.-R. Sack, Minimum Decompositions of Polygonal Objects, in *Computational Geometry,* G.T. Toussaint, Ed., North Holland, (1985).
[14] Lee, D.T. and F.P Preparata, Location of a Point in a Planar Subdivision and its Applications, *SIAM Journal on Computing,* Vol. 6, No.3, 594-606 (1977).
[15] Lee, D.T., Visibility of a Simple a Polygon, *Computer Vision, Graphics and Image Processing,* Vol. 22, No. 2, 207-221 (1983).
[16] Lee, D.T. and A.K. Lin, Computational Complexity of Art Gallery Problems, *IEEE Transactions Information Theory,* Vol. IT-32, 276-282 (1986).
[17] Lubiw, A., Decomposing Polygonal Regions into Convex Quadrilaterals, *Proceedings of the Symposium on Computational Geometry,* 97-106 (1985).
[18] Manilla, H. and D. Wood, A simple Proof of the Rectilinear Art Gallery Theorem, Tech. Report C-1984-16, Department of Computer Science, University of Helsinki, Finland (1984).
[19] O'Rourke, J., An Alternate Proof of the Rectilinear Art Gallery Theorem, *Journal of Geometry,* Vol. 21, 118-130 (1983).
[20] O'Rourke, *J. Art Gallery Theorems and Algorithms,* Oxford University Press (1987).
[21] Sack, J.-R. and G.T. Toussaint, A Linear-Time Algorithm for Decomposing Rectilinear Star-Shaped Polygons into Convex Quadrilaterals, *Proceedings of the 19th Annual Conference on Communication, Control, and Computing,* 21-30 (1981).
[22] Sack, J.-R., An O(n log n) Algorithm for Decomposing Simple Rectilinear Polygons into Convex Quadrilaterals, *Proceedings of the 20th Annual Conference on Communication, Control, and Computing,* 64-74 (1982).
[23] Sack, J.-R., A Simple Hidden-Line Algorithm for Rectilinear Polygons, *Proceedings of the 21st Allerton Conference on Communication, Control and Computing,* Urbana-Champaign, 437-446 (1983).
[24] Sack, J.-R., Rectilinear Computational Geometry, Tech. Rept. SCS-TR-54, School of Computer Science, Carleton University, Ottawa, Canada (1984).
[25] Tarjan, R. and C.J. Van Wyck, An O(n loglog n) Algorithm for Triangulating Simple Polygons, preprint (submitted to *SIAM J. Computing)* (1986).

Computational Morphology
G.T. Toussaint (editor)

REALIZABILITY OF POLYHEDRONS FROM LINE DRAWINGS†

Kokichi SUGIHARA

Department of Mathematical Engineering and Information Physics
Faculty of Engineering, University of Tokyo
Hongo 7-3-1, Bunkyo-ku, Tokyo 113, Japan

The paper gives a tutorial survey of the realizability problem of polyhedrons from line drawings. The realizability problem is to judge whether a given line drawing represents a polyhedron. The problem includes many variations depending on assumptions about the object worlds, on the view point or the view direction, on the drawing rules, etc. These variations are classified, and main results in each class of problems are summarized.

1. INTRODUCTION

The realizability problem we consider in this paper is the problem of judging whether a three-dimensional polyhedral object is realizable from a given two-dimensional line drawing. The problem arises in various fields such as projective geometry, graphic science, and scene analysis.

Mathematical structures of line drawings have long been studied in projective geometry and descriptive geometry (Hohenberg [1] and Japan Society for Graphic Science [2]). The main problem there is the description problem, that is, how to describe three-dimensional structures in terms of two-dimensional figures. The converse problem, how to reconstruct three-dimensional structures from two-dimensional line drawings, drew attention only when the digital computer became capable of handling pictorial data (Duda and Hart [3], Winston [4] and Ballard and Brown [5]). Nevertheless the problem includes many variations. This is probably because the problem arises in several different fields in mathematics and engineering almost simultaneously and independently.

In this paper, we first give a framework to classify various types of realizability problems in the next section, and then summarize important results in the subsequent sections.

2. REALIZABILITY PROBLEMS

2.1. Introductory Consideration

A *polyhedron* is a solid object bounded by a finite number of planar *faces*. It is not necessarily convex; it may be concave or even have holes. Line segments shared by two faces are called *edges*, and points where three or more faces meet are called *vertices*. A *line drawing* (or *picture* in short) is a two-dimensional figure composed of a finite number of straight line segments. Here we consider the problem of judging whether a given line drawing represents a polyhedron.

† The work is partly supported by the Grant-in-Aid for Scientific Research of the Ministry of Education, Science, and Culture of Japan (Grant No. 62580017).

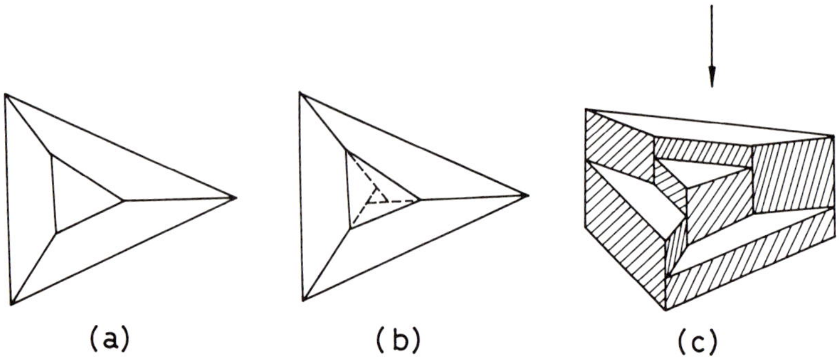

FIGURE 1. Intuitive interpretation and tricky interpretation of a line drawing.

The most naive type of the realization problem is the following.

Naive Problem. Given a line drawing, judge whether a polyhedron is realizable from the line drawing.

However, this is not a well-defined question. For example, consider a line drawing shown in Figure 1(a). How can one answer if one is asked whether a polyhedron is realizable from this line drawing? One may say yes, because it can usually be used as an illustration of a truncated pyramid seen from above. In a mathematical sense, however, it cannot be a picture of any truncated pyramid; indeed the three side faces of a truncated pyramid should have a common point of intersection in a space (when extended), and hence the three lines corresponding to the pairwise intersections of the three faces should meet at a common point on the picture plane, but they do not as shown in (b). So it seems that one should answer that a polyhedron is not realizable from this line drawing.

We can, however, object to this answer. Indeed the line drawing cannot be a picture of any truncated pyramid, but can be a picture of other kinds of polyhedrons. The object shown in (c) is an example of a polyhedron that can yield the line drawing in (a). This object has one bottom face and four top faces with different heights; the other faces, which are indicated by shaded areas in (c), are all vertical. Projecting the object orthographically in the vertical direction (the direction indicated by the arrow in (c)), we get the line drawing in (a). Thus, the picture can represent a polyhedron. This answer, however, is not very useful because we are not interested in such tricky objects and tricky situations.

In order to avoid "pathological" answers, we have to introduce several assumptions on objects, on view points, on drawing rules, etc.

2.2. Assumptions on Objects

First, in order to avoid unusual polyhedrons we make the next assumption.

Assumption O1. Polyhedrons are bounded and homogeneously three-dimensional.

Since the polyhedrons are bounded, infinite objects such as a half space are excluded. Since they are homogeneously three-dimensional, lower-dimensional objects such as a sheet of paper are also excluded.

It is often useful to restrict the objects to a certain subclass of the polyhedrons. Some typical examples of the restriction are the following.

Assumption O2. Polyhedrons are topologically equivalent to a ball.
Assumption O3. Polyhedrons are convex.

A vertex is called *trihedral* if it is on exactly three faces, and a polyhedron is called *trihedral* if all the vertices are trihedral. A vertex is called *rectangular* if all the faces meeting at the vertex are mutually perpendicular, and a polyhedron is called *rectangular* if all the vertices are rectangular. The following restrictions are also important.

Assumption O4. Polyhedrons are trihedral.
Assumption O5. Polyhedrons are rectangular.

Note that Assumption O5 does not imply Assumption O4. The object shown in Figure 2 is an example that is rectangular but not trihedral; six faces meet at the central vertex.

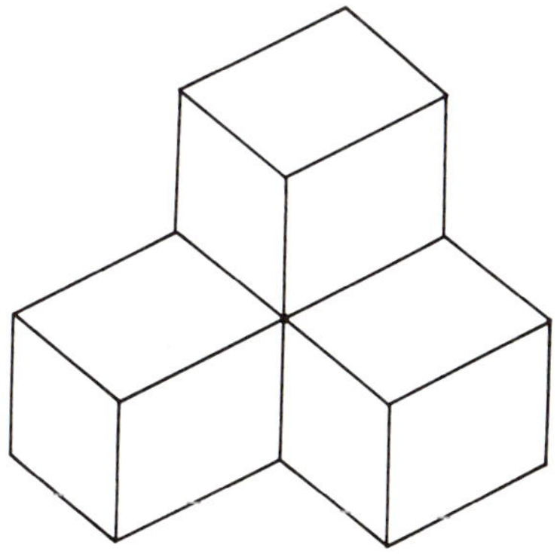

FIGURE 2. Polyhedron that is rectangular but not trihedral.

2.3. Assumptions on Points of View

Let P be a polyhedron fixed in a three-dimensional space, e be a point outside P, and s be a plane that does not contain e. Projecting P onto the plane s with respect to the center of projection at e, we get a picture of P, as shown in Figure 3. The picture can be regarded as the image of the polyhedron P seen by the viewer whose eye is centered at e and whose retinal plane is s. The point e is referred to as the *view point*. In the projection, a point p on P corresponds to the point of intersection of the line pe and the plane s (where pe denotes the line passing through p and e).

We can also imagine the case in which the view point e recedes infinitely far in a certain direction; this direction is called the *view direction*. In this case the line pe is parallel to the direction of the projection, so that the picture of P is defined unless the view direction is parallel to s.

The projection is called *perspective* or *central* if the view point e is at a finite distance, and *parallel* if it is at infinity. The parallel projection is called *orthographic* if the view direction is perpendicular to the picture plane s, and *oblique* otherwise. In what follows we often use the terminology for perspective projection to represent situations for both perspective and parallel projections. For example, if we say that a plane passes through the view point e, it means in the case of parallel projection that the plane is parallel to the view direction.

The following assumptions are useful for prohibiting some tricky answers.

Assumption W1. No face of a polyhedron is coplanar with the view point.
Assumption W2. No pair of edges of a polyhedron is coplanar with the view point unless the two edges are collinear in the space.
Assumption W3. No pair of a vertex and an edge of a polyhedron is coplanar with the view point unless the vertex is one of the two terminal points of the edge.
Assumption W4. No pair of vertices of a polyhedron is collinear with the view point.

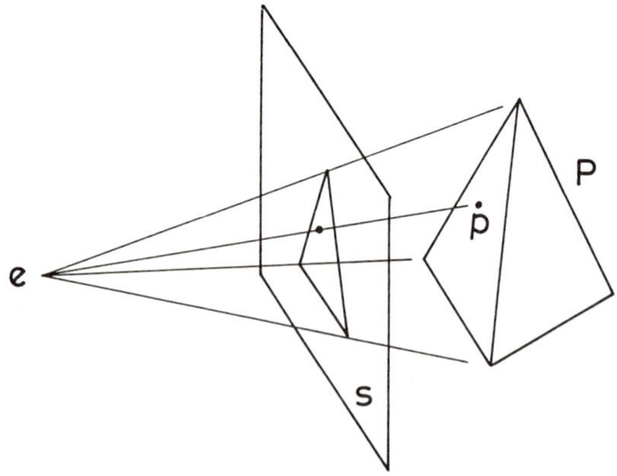

FIGURE 3. An object, the view point, and the picture.

Assumption W1 implies that every face of a polyhedron corresponds to some region of the picture with nonzero area. Assumption W2 implies that neither distinct edges accidentally overlap in the picture nor noncollinear edges become collinear in the picture. Assumption W3 implies that a vertex does not lie on an edge in the picture unless the vertex is originally a terminal point of the edge. Assumption W4 implies that two distinct vertices never fall at the same point on the picture plane. Note that Assumption W2 implies Assumption W1, and that Assumption W3 implies both Assumptions W2 and W4 (and hence implies Assumption W1, too). These types of assumptions are often called *nonaccidentalness assumptions* (Marr [6]). The view point is said to be in *general position* if all, or some, of the above four assumptions are true; different authors use different definitions (see Huffman [7], Clowes [8], Waltz [9], Sugihara [10], and Kanade [11]).

The situation shown in Figure 1(c), i.e., the mutual relationship between the object and the view direction shown there, violates Assumptions W1, W2, W3, and W4; consequently, if we choose at least one of these four assumptions, we can avoid the tricky answer.

2.4. Assumptions on Drawing Rules

One more class of assumptions is necessary to clear up ambiguity in the realizability problem. They are about drawing rules. First, we take the next assumption.

Assumption D1. Edges only are drawn in line drawings.

Hence, even if the object surface has texture, scribbles, cracks, or shadows, they are not drawn. Each line segment in the picture corresponds to some edge of the object.

Next, we have to make it clear how to treat hidden edges. The following three alternative assumptions are most common and important.

Assumption D2. Visible edges only are drawn.
Assumption D3. Visible edges are drawn with solid lines, and invisible edges are drawn with broken lines.
Assumption D4. All edges, visible and invisible, are drawn with solid lines.

Assumption D2 is useful for image analysis, and Assumption D3 for interpretation of engineering drawings. Assumption D4 seems less realistic, but more fundamental for the study of the mathematical structure of line drawings.

2.5. Other Variations

We have listed many assumptions which seem necessary or useful to make the realizability problem senseful. We usually adopt some subset of these assumption set. Different subsets define different problems. In what follows we consider the following type of problems.

Problem with Assumption (PA). Given a line drawing and a set of assumptions, judge whether a polyhedron is realizable from the line drawing.

Besides the choice of the subset of assumptions, however, there are still other factors that can generate variations of the problems.

The first factor is the view point. When an object is projected onto the picture plane, a certain view point is used. The realizability in general depends on whether the view point is given or not. If the view point relative to the picture plane is given, we have to judge whether the picture is the projection of some polyhedron with respect to the given view point. If it is not given, on the other hand, we have to search for both a polyhedron and a view point that together give the picture. Thus, we get another type of a realizability problem.

Problem with the view point (PV). Given a line drawing and the view point, judge whether a polyhedron is realizable from the line drawing with respect to the view point.

The dependence of the realizability problem on the view point will be discussed in Sections 3 and 8.

The second factor is whether additional knowledge about the structure of the object is given or not. For example, to ask whether the line drawing in Figure 1(a) represents any polyhedron is different from to ask whether it represents a truncated pyramid. Thus, we can consider the following type of a realizability problem.

Problem with additional knowledge (PK). Given a line drawing and some additional knowledge about the structure of a polyhedron, judge whether a polyhedron with the given structure is realizable from the line drawing.

There are many different ways of giving additional knowledge. A typical example of such knowledge is given by labels assigned to lines, as will be seen in Section 4; a problem in which additional knowledge about the structure of the object is not given is usually decomposed into a set of subproblems of the type PK.

The third factor is whether the realizability is judged strictly or flexibly. As we have seen, the line drawing in Figure 1(a) is not a correct projection of a truncated pyramid. This kind of incorrectness, however, usually arises in practical data; digitization errors are inevitable when picture data are given to a computer. If we want to construct a computational mechanism for interpreting engineering drawings or analyzing photographic images, the realizability should be judged "flexibly" in the sense that pictures are regarded as being "practically correct" if incorrectness is only due to small errors in vertex positions. Thus, another type of a problem is the following.

Problem requiring flexibility (PF). Given a line drawing, judge whether it can be regarded (in some practical sense) as a picture of a polyhedron.

In this way we shall distinguish the problem types by the symbols PA, PV, PK, and PF. We shall use the symbols also in combination. For example, PAV stands for the type of a problem in which a set of assumptions and the view point are given, PAVK stands for the type of a problem in which a set of assumptions, the view point, and additional knowledge about the structure of the object are given, etc.

3. PERSPECTIVE, OBLIQUE, OR ORTHOGRAPHIC

Now we see how the realizability problem is affected by the information about the view point. Here we fix a line drawing to the *xy* plane, and consider various view points and view directions.

First let us consider parallel projection. Let P be a polyhedron fixed in the space, $\mathbf{v} = (v_x, v_y, v_z)$ be the three-dimensional vector that is not parallel to the *xy* plane, and let D be a line drawing generated by the parallel projection of P onto the *xy* plane with respect to the view direction $\mathbf{v}$, as shown in Figure 4. Thus, a polyhedron is realizable from the line drawing D with respect to the view direction $\mathbf{v}$. Now, let $\mathbf{v}' = (v_x', v_y', v_z')$ be another view direction, and consider the question: "Is a polyhedron realizable from D with respect to the view direction $\mathbf{v}'$?" In principle, the answer should depend on the assumption which we make in the problem. In most cases, however, it does not depend as the following shows.

Let us consider the transformation of points from (x, y, z) to (x', y', z') defined by

$$(1) \qquad \begin{bmatrix} x' \\ y' \\ z' \end{bmatrix} = \begin{bmatrix} 1 & 0 & (\alpha v_x' - v_x)/v_z \\ 0 & 1 & (\alpha v_y' - v_y)/v_z \\ 0 & 0 & \alpha v_z'/v_z \end{bmatrix} \begin{bmatrix} x \\ y \\ z \end{bmatrix},$$

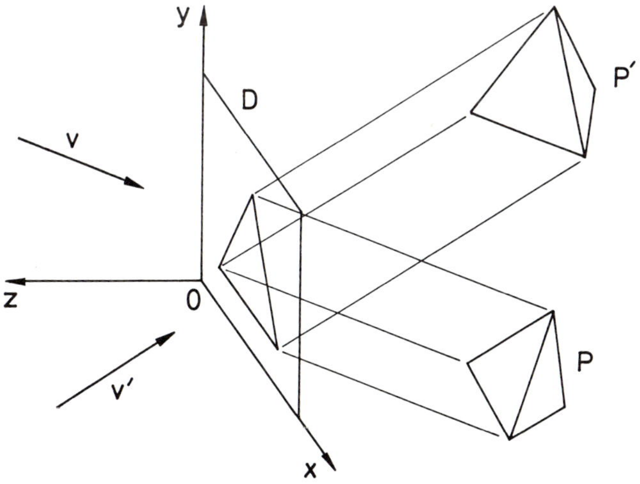

FIGURE 4. Parallel projection.

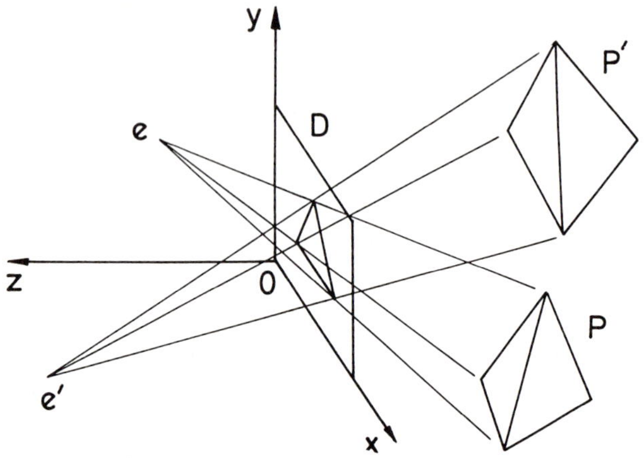

FIGURE 5. Perspective projection.

where α is a nonzero parameter. Let P' be the set of all points (x', y', z') obtained by the transformation when (x, y, z) moves in P. The transformation is affine, and consequently, if (x, y, z) moves on a line, (x', y', z') also moves on a line and if (x, y, z) moves on a plane, (x', y', z') also moves on a plane. Thus, P' is a polyhedron. It is obvious that if $z = 0$, then $x' = x$, $y' = y$, and $z' = 0$; thus any point on the xy plane remains unchanged when the transformation is applied to the space. Moreover, the vector $\mathbf{v} = (v_x, v_y, v_z)$ is mapped to $\alpha v'$ $= (\alpha v_x', \alpha v_y', \alpha v_z')$ by the transformation. These observations together tell us that if D is the projection of P with respect to the view direction $\mathbf{v}$, D is also the projection of P' with respect to the view direction $\mathbf{v}'$.

Affine transformations preserve affine properties such as collinearity and coplanarity, but does not preserve metric properties such as distances and angles. Hence, among Assumptions O2, O3, O4, and O5, only Assumption O5 is disturbed by affine transformations. Thus, if the object world is restricted by Assumption O2, O3, or O4 and only parallel projections (i.e., the orthographic projection and the oblique projections) are considered, the realizability problem does not depend on the view direction.

Next, let us consider perspective projections. Let P be a polyhedron, $\mathbf{e} = (e_x, e_y, e_z)$ be a point outside the xy plane, and D be the line drawing obtained by the perspective projection of P onto the xy plane with respect to the view point $\mathbf{e}$, as shown in Figure 5. Now, $\mathbf{e}' =$ (e_x', e_y', e_z') be another point outside the xy plane, and consider the question: "Is a polyhedron realizable from D with respect to the view point $\mathbf{e}'$?"

Let us consider the transformation from (x, y, z) to (x', y', z') defined by

$$(2)\qquad \begin{cases} x' = \dfrac{e_z x + (\alpha e_z e_x' + e_x' - e_x)z}{e_z(\alpha z + 1)}, \\ y' = \dfrac{e_z y + (\alpha e_z e_y' + e_y' - e_y)z}{e_z(\alpha z + 1)}, \\ z' = \dfrac{(\alpha e_z + 1)e_z' z}{e_z(\alpha z + 1)}, \end{cases}$$

where α is a nonzero parameter. The transformation maps $\mathbf{e} = (e_x, e_y, e_z)$ to $\mathbf{e}' = (e_x', e_y', e_z')$, and for any point on the xy plane, the transformation maps the point to itself. Moreover, a direct calculation shows that if (x, y, z) moves on a line, then (x', y', z') defined by the transformation (2) also moves on a line (for example, if we assume the parametric form of the line as $x = At + D$, $y = Bt + E$, $z = Ct + F$, then we get a parametric form $x' = A't' + D'$, $y' = B't' + E'$, $z' = C't' + F'$, where $t' = t/(e_z(\alpha Ct + \alpha F + 1))$ and A', B', C', D', E', F' are some constants depending on A, B, C, D, E, F, $\mathbf{e}$, $\mathbf{e}'$). Hence, P' is also a polyhedron, and D is a perspective projection of P' with respect to the view point $\mathbf{e}'$ (Figure 5). Therefore, in the case of perspective projection too, if a polyhedron belonging to the object world under consideration is mapped by the transformation (2) to a polyhedron in the same object world, then the realizability does not depend on the view point. In our examples of assumptions, the problem is independent from the view point in Assumptions O2, O3, and O4, whereas it is dependent in Assumption O5.

Finally, let us consider the relationship between the perspective projections and the parallel projections. Let D be the line drawing obtained by the perspective projection of a polyhedron P onto the xy plane with respect to the view point $\mathbf{e} = (e_x, e_y, e_z)$, and let us consider the next transformation.

$$(3)\qquad \begin{cases} x' = e_x - \dfrac{e_z(x - e_x)}{z - e_z}, \\ y' = e_y - \dfrac{e_z(y - e_y)}{z - e_z}, \\ z' = -\dfrac{e_z z}{z - e_z}. \end{cases}$$

The transformation (3) maps the line drawing D to itself, the view point e to infinity, and the polyhedron P to another polyhedron, say P'. Moreover, the line drawing D is also obtained as the orthographic projection of P' onto the xy plane. Thus, if D is a perspective projection of a polyhedron, it is also an orthographic projection of some other polyhedron.

Note that the transformation (1) is an affine transformation, whereas the transformations (2) and (3) are projective ones. Hence, in (2) and (3) only projective properties such as concurrency, collinearity, and coplanarity of geometric objects are preserved, but in (1) affine properties such as parallelism of lines and/or planes and ratios along the same line are also preserved. Metric properties such as lengths and angles are not preserved by these transformations. Therefore, if the object world is specified only by projective properties, the realizability problem does not depend on the view point or the view direction, and if it is specified by affine properties, the realizability problem under the parallel projection assumption does not depend on the view direction. If the object world is specified by metric properties, on the other hand, the realizability problem depends critically on the view point or the view direction. Among Assumptions O1, ..., O5, only O5 is metric and all the other are projective. Hence, in particular we get the next theorem.

Theorem 1. With Assumption O2, O3, or O4, a problem of the type PAV is equivalent to the corresponding problem of the type PA.

With Assumption O5, on the contrary, the realizability problem gives different answers for different view points or view directions.

An example of affine properties is parallelism. Hence, the realizability problem with additional knowledge that the object is a parallelepiped, for example, does not depend on the view direction if we consider parallel projections, but depends on the view point if we consider perspective projections.

4. LABELING SCHEME

Obviously a problem of the type PAK is a subproblem of the corresponding problem of the type PA. Here we shall see a method for decomposing a type-PA problem into type-PAK subproblems. The method was first developed by Huffman [7] and Clowes [8] independently. The following notations are due to Huffman [7].

Given a line drawing *D*, let us consider a type-PA problem with Assumptions O1, O4, W3, D1, and D2, that is, a realizability problem of a hidden-line-eliminated line drawing in the trihedral object world with the nonaccidentalness assumption (recall Assumption W3 implies Assumptions W1, W2, and W4). Then, each line in the picture corresponds to exactly one edge of the object. We classify the lines into three categories and assign labels in the following way. A line corresponding to a convex edge whose both sides are visible is labeled with "+", a line corresponding to a concave edge is labeled with "-", and a line corresponding to a convex edge one of whose side faces is hidden by the other is labeled with an arrow; the direction of the arrow is chosen so that the visible side face is to the right of the arrow.

Huffman observed that with Assumptions O1, O4, W3, D1, and D2, only the twelve types of vertices listed in Figure 6 can appear in line drawings of polyhedrons. Hence, we can extract probable candidates for the spatial structure of the object by finding assignments of the three kinds of labels to the picture that are consistent with the vertex list in Figure 6.

In Figure 7, the pictures in (a) and (b) admit consistent assignments of labels; the labeling in (a) corresponds to the interpretation of the picture as a triangular pyramid, and that in (b) as a truncated pyramid. The picture in (c), on the other hand, has no consistent way of labeling, and hence we can conclude that a polyhedron is not realizable from the picture (the top face and the left front face share two noncollinear edges, which is impossible unless some

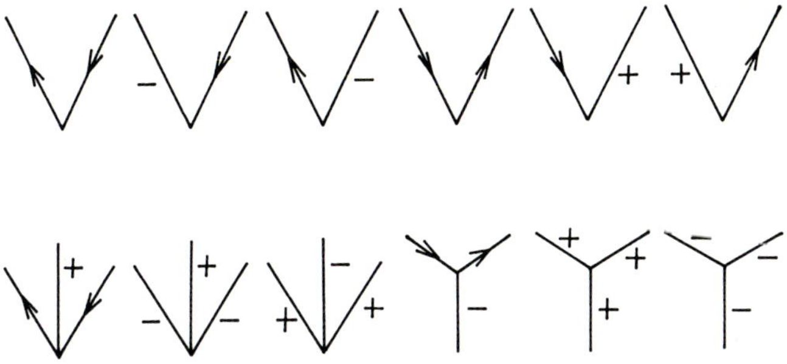

FIGURE 6. List of possible vertices in the trihedral object world (adapted from Huffman [7]).

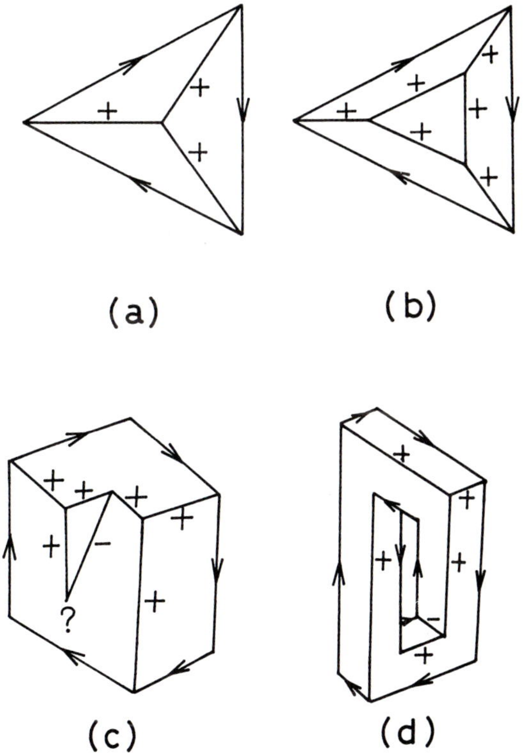

FIGURE 7. Consistent labelling.

face is curved). The picture in (d) has a consistent way of labeling, but it does not correspond to any polyhedron. Thus, for a picture to admit a consistent labeling is a necessary, but not sufficient, condition for the picture to represent a polyhedron correctly. A labeling that is consistent with the list in Figure 6 gives merely a candidate for the spatial structure of the object.

A consistent labeling can be regarded as a representation of knowledge about the spatial structure of the object. Consequently, the realizability problem of a labeled line drawing can be regarded as a type-PK problem. In general, a consistent way of labeling, if exists, is not unique; a line drawing often admits two or more consistent ways of labeling. Thus, the vertex list in Figure 6 can decompose a type-PA problem into a set of type-PAK problems.

If some of the assumptions are replaced with others, the list of possible vertices should be changed. Many other lists are constructed. Sugihara [10] gives the list for the case in which Assumption D2 is replaced with D3, that is, the list for hidden-edge-drawn pictures (see also Sanker [12]). Waltz [9] gives a huge list for pictures with shadows and cracks in which accidental alignments of lines and vertices are partially allowed, Kanade [11] gives a list for paper-made objects, Nakatani and Kitahashi [13] and Kanatani [14] give lists for rectangular objects, etc. Lists of possible vertices are used also for interpreting pictures of quadric-surface objects (Turner [15]). Moreover, it is possible to automatically construct the list of vertices in a given object world (Sugihara [16]).

As for time efficiency for finding consistent labelings, some nontrivial techniques are necessary because direct search leads to combinatorial explosions. Suppose that the picture has l lines. Since in Huffman's scheme each line can have one of the four different labels (note that an arrow can be assigned in one of the two mutually opposite directions), the number of all combinations of assigning labels amounts to 4^l, which is a tremendous number even when $l = 15$ or so. Sometimes inconsistency in labeling can be recognized very locally, and hence a large part of the search tree can be pruned on early stages of the search process. A typical method for pruning is given by the technique called *constraint propagation*; see Montanari [17], Waltz [9], Rosenfeld *et al.* [18], Mackworth [19], McGergor [20], Haralick and Shapiro [21, 22], and Nudel [23].

However, recently the problem of finding consistent assignments of labels was proved to be NP-complete by Kirousis and Papadimitriou [24]; hence it is almost impossible to get a method that works always efficiently.

5. GRADIENT SPACE AND RECIPROCAL FIGURES

5.1. Perpendicularity Condition

Here we consider type-PK problems (that is, problems where some additional knowledge about the structures of the objects are given), and present a necessary condition for a line drawing to be a correct projection of the object. As we have seen in Section 3, the realizability does not depend on the view point or the view direction unless the object world is restricted by metric properties. Hence, we concentrate on the orthographic projection of objects onto the xy plane.

Let P be a polyhedron fixed in the space, and f_j be the jth face of P. Let

$$a_jx + b_jy + z + c_j = 0 \tag{4}$$

denote the planar surface on which f_j lies. (This equation cannot represent a plane that is parallel to the z axis, but there is no problem if we take Assumption W1.) The equation has three parameters a_j, b_j, and c_j; a_j and b_j specify the orientation of the surface whereas c_j represents the degree of freedom due to translation. The ordered pair $G_j = (a_j, b_j)$ is called the *gradient* of the surface. If a_j and b_j are considered as the horizontal and vertical coordinates, the gradient G_j can be regarded as a point on a two-dimensional space with the (a, b) coordinate system. This space is called the gradient space. Thus, each family of mutually parallel surfaces corresponds to a point in the gradient space.

The use of the gradient space for the realizability problem is based on the following remarkable property. Let f_1 and f_2 be two faces that share a common edge. The common edge is represented by a pair of equations obtained from eq. (4) by putting $j = 1$ and $j = 2$. The orthographic projection, say l, of the edge onto the xy plane is represented by the equation that is obtained when z is eliminated from the pair of equations:

$$(a_1 - a_2)x + (b_1 - b_2)y + (c_1 - c_2) = 0. \tag{5}$$

This equation implies that the line l is perpendicular to the vector $(a_1 - a_2, b_1 - b_2)$. In other words, if we superimpose the gradient space upon the picture plane in such a way that the x and y axes are respectively parallel to the a and b axes, then the line connecting a pair of the gradients of faces that share a common edge is perpendicular to the image of the edge.

In type-PK problems we are given information about which faces share common edges. Consequently the realizability can be partly checked by the above perpendicularity condition.

Theorem 2. For a type-PK problem, a polyhedron is realizable only when the faces can be mapped to distinct points in the picture plane in such a way that, if a pair of faces share a common edge, the line connecting the corresponding pair of points is perpendicular to the image of the edge.

The inconsistency of the interpretation of the line drawing in Figure 1(a) as a picture of a truncated pyramid can be recognized by this condition, as shown in Figure 8(a). Since the condition is invariant under translation and scaling of the figure, we can plot the first point G_1 arbitrarily and the second point G_2 in an arbitrary distance from G_1. If we proceed plotting G_1, G_2, G_3 in this order using the condition stated in Theorem 2, we find that the last point G_4 cannot be plotted. Thus, we can recognize the inconsistency in interpreting the picture as a truncated pyramid. Theorem 2 or something similar has been used for the recognition of inconsistency in labeled line drawings by Huffman [7, 25, 26, 27], Mackworth [28], Whiteley [29], Kanade [11, 30], and Draper [31].

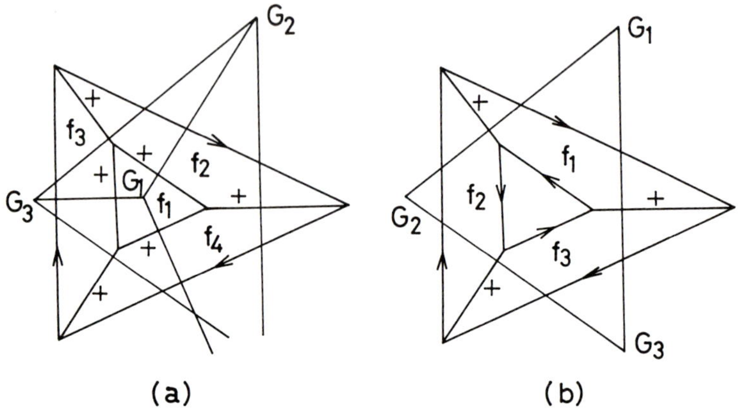

FIGURE 8. Perpendicularity condition.

However, the condition is not sufficient. For example, the labeled line drawing in Figure 8(b) represents the interpretation of the picture as an object that is obtained from a truncated pyramid by digging a hole so that a triangular top face is replaced with a hole penetrating the object. This interpretation is likewise incorrect, but we can plot the gradients of all the three visible faces without violating the perpendicularity condition, as shown in (b). Thus, the condition is necessary but not sufficient even if there is no overlap among visible faces.

Huffman [27] proved that a slightly modified version of the perpendicularity condition is necessary and sufficient for an isolated vertex and the faces around it to be realizable. Recently Aurenhammer [32] showed, as a corollary of a more general theorem, that the perpendicularity condition is necessary and sufficient for a tiling of a plane with a finite number of convex polygons to be a projection of a convex polyhedron (see also Ash *et al.* [33]).

It should be noted that, though we restrict our consideration to the orthographic projections, the perpendicularity condition is valid also for oblique and perspective projections. This is because the realizability does not depend on the view point (recall Theorem 1). Note, however, that in the case of the oblique or perspective projection, the plotted points $G_j = (a_j, b_j)$ do not mean "gradients" of the faces.

5.2. Reciprocity Condition

If hidden edges are also drawn in the line drawing, the perpendicularity condition stated in Theorem 2 can be replaced with a stronger condition, which we see in this subsection.

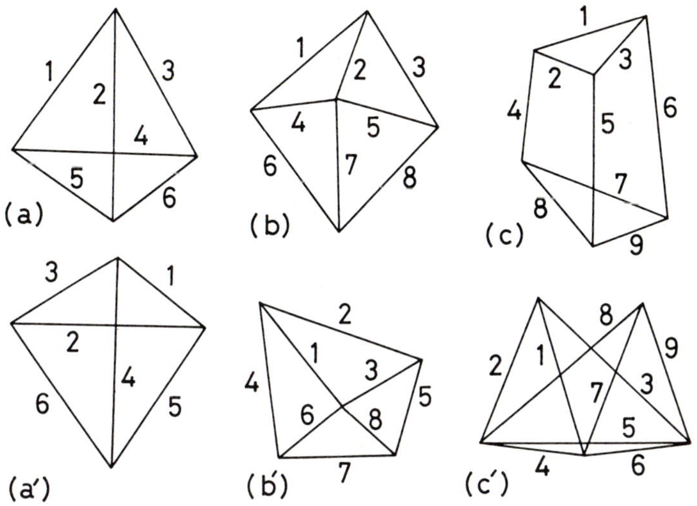

FIGURE 9. Reciprocal pairs of diagrams.

Let $G = (V, E)$ be an undirected graph consisting of the node set V and the arc set E. A *diagram* is a triple $D = (V, E, h)$ where (V, E) is a graph and h is a mapping from V to $\mathbf{R}^2$ ($\mathbf{R}^n$ denotes an n-dimensional Euclidean space). The diagram can be thought of as a picture of the graph G drawn on a plane in such a way that the node $v \in V$ is plotted at $h(v)$ and arcs are drawn by straight line segments connecting the corresponding nodes. Diagrams defined as above are very similar to line drawings. There is, however, difference. In line drawings it is not clear whether a point at which two or more noncollinear lines meet is an image of some vertex or an accidental crossing of spatially apart two edges. In the diagram, on the other hand, the nodes are given by V explicitly, and hence they can be distinguished from accidental crossings.

Let $D = (V, E, h)$ and $D' = (V', E', h')$ be two diagrams. D and D' are said to be *reciprocal* if there is a one-to-one correspondence between the arc sets E and E', so that corresponding pairs of arcs are perpendicular, and corresponding arcs converging to a node in one diagram form a closed path in the other.

Figure 9 gives three pairs of diagrams that are mutually reciprocal. In each pair the corresponding arcs are labeled with the same number. We see, for example, that the six pairs of arcs in (a) and (a') are mutually perpendicular, and that the three arcs 1, 2, and 3 in (a) are converging to a node whereas the corresponding three arcs in (a') form a closed path.

Let P be a polyhedron fixed in the space, and V, E, and F be the set of vertices, that of edges, and that of faces, respectively, of P. Regarding V and E as the node set and the arc set, we get a graph, which we call the *vertex-edge graph* of P and denote by $VG(P) = (V, E)$.

Furthermore, regarding F and E as the node set and the arc set (where an arc connects two nodes if the corresponding two faces share a common edge), we get another graph, which we call the *face-edge graph* of P and denote by $FG(P) = (F, E)$. Let h_V be a mapping from V to $\mathbf{R}^2$ such that $h_V(v)$ is the orthographic projection of the vertex v onto the xy plane. Then, $VD(P) = (V, E, h_V)$ is a diagram, which we call the *vertex-edge diagram* of P. Let h_F be a mapping from F to $\mathbf{R}^2$ such that $h_F(f)$ is the gradient of the face f. Then, $FD(P) = (F, E, h_F)$ is a diagram, which we call the *face-edge diagram* of P.

The vertex-edge diagram can be thought of as the picture of P in which hidden lines are also drawn by solid lines (that is, a picture in Assumption D3). The face-edge diagram, on the other hand, can be regarded as the figure constructed in the gradient space using the constraint stated in Theorem 2. This observation leads to the next theorem (Maxwell [34, 35] and Cremona [36]; see also Whiteley [29], and Sugihara [37]).

Theorem 3. For any polyhedron P fixed in the space, the vertex-edge diagram and the face-edge diagram are reciprocal.

It follows from the theorem that for a type-PK problem with Assumption D3 or D4 (i.e., hidden lines are drawn in the line drawing), a polyhedron is realizable only when the line drawing has a reciprocal diagram. In Figure 9, the pictures in (a), (b), and (c) are vertex-edge diagrams of a tetrahedron, a quadrilateral pyramid, and a triangular prims, and indeed they have reciprocal diagrams as shown in (a'), (b'), and (c').

However, the reciprocity condition is not sufficient, either, for a picture to represent a polyhedron. The line drawing in Figure 9(c') has a reciprocal diagram as shown in (c), but obviously does not represent any polyhedron.

Crapo and Whiteley [38] generalizes the notion of a polyhedron that is topologically equivalent to a sphere to an "oriented spherical polyhedron," in which the surface may penetrate itself, and shows that, if the object world is restricted to the "oriented spherical polyhedrons," for a line drawing to have a reciprocal diagram is a necessary and sufficient condition for a polyhedron to be realizable.

6. LINEAR-ALGEBRAIC APPROACH

6.1. Realizability of an Incidence Structure

One fundamental property of the realizability problem stems from the planarity of faces. Here we concentrate on this property, and consider a rather abstract but fundamental problem. The result can be used for the representation of a necessary and sufficient condition for the realizability, as will be shown in the next subsection.

An *incidence structure* is a triple $I = (V, F, R)$, where $V = \{v_1, ..., v_n\}$ and $F = \{f_1, ..., f_m\}$ are mutually disjoint finite sets and R is a subset of $V \times F$. The elements of V are called *points*, those of F *surfaces*, and those of R *incidence pairs*.

A *picture* of I, denoted by $I(\mathbf{p})$, is the incidence structure I together with a $2n$-dimensional real vector $\mathbf{p} = (x_1, y_1, ..., x_n, y_n) \in \mathbf{R}^{2n}$. We refer to the points (x_i, y_i) $(i = 1, ..., n)$ as the vertices of $I(\mathbf{p})$.

A realization of $I(\mathbf{p})$ is an $(n+3m)$-dimensional real vector $\mathbf{w} = (z_1, ..., z_n, a_1, b_1, c_1, ..., a_m, b_m, c_m) \in \mathbf{R}^{n+3m}$ such that

(6) $$a_j x_i + b_j y_i + z_i + c_j = 0$$

for any (v_i, f_j) in R. The realization can be considered as a collection of n points and m planar surfaces (extended infinitely) in the three-dimensional space, where the points are represented by Cartesian coordinates (x_i, y_i, z_i) and the surfaces by the equations

(7) $$a_j x + b_j y + z + c_j = 0$$

such that the orthographic projection of the points onto the xy plane coincides with the vertices of $I(\mathbf{p})$ and that, for any (v_i, f_j) in R, the ith point is on the jth surface. The realization $\mathbf{w}$ is said to be *nondegenerate* if $(a_i, b_i, c_i) \neq (a_j, b_j, c_j)$ for any $1 \le i < j \le m$. We say that I is *realizable* from $I(\mathbf{p})$ if $I(\mathbf{p})$ has a nondegenerate realization.

Intuitively, the realizability of an incidence structure from $I(\mathbf{p})$ is an abstraction of the realizability of a polyhedron. An incidence structure is obtained from a polyhedron by extending faces infinitely and ignoring which side of a face is occupied with material.

Suppose that a picture $I(\mathbf{p})$ is given. Then, $x_1, y_1, ..., x_n, y_n$ are known, and hence eq. (6) is linear in z_i, a_j, b_j, and c_j. Collecting the equations for all elements in R, we get the system of linear equations

(8) $$A(I, \mathbf{p})\mathbf{w}^t = \mathbf{0},$$

where $A(I, \mathbf{p})$ is a constant matrix of size $|R| \times (n + 3m)$ and t denotes the transposition. A vector $\mathbf{w}$ is a realization of $I(\mathbf{p})$ if and only if it is a solution of eq. (8). Thus, the realizations of $I(\mathbf{p})$ form a linear subspace of $\mathbf{R}^{n+3m}$, and its dimension is $n + 3m - \mathrm{rank}(A(I, \mathbf{p}))$. This dimension is called the *degree of freedom* of $I(\mathbf{p})$. For a realizable picture, the degree of freedom represents how much different objects can yield the picture.

Remark: It may seem that to have a reciprocal diagram is necessary and sufficient for an incidence structure to be realizable, but it is not true. A counterexample is again given in Figure 8(b) when we consider the figure as a picture $I(\mathbf{p})$ of the incidence structure composed of the six vertices and the three faces.

6.2. A Necessary and Sufficient Condition for Realizability

Let us consider the realizability problem for a labeled line drawing, that is, a type-PK problem where additional knowledge about the structure of the object is given by the labels. Let D be a labeled line drawing in which visible edges only are drawn. The labels tell us which vertices are on which faces, and hence the incidence structure of the object can be extracted explicitly. Consequently the system of linear equations, eq. (8), is constructed.

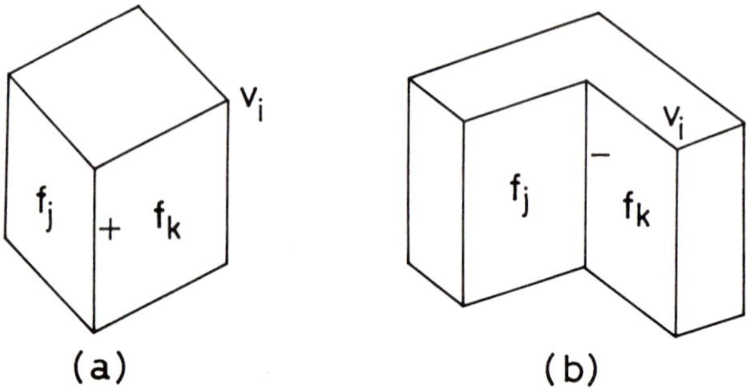

FIGURE 10. Cues for relative depth.

Eq. (8) represents only the constraints placed by the planarity of the faces, whereas the labels in the picture constrain also some other properties of the object; for example, the property as to which side of a face is filled with material or whether an edge forms a ridge or a valley. These constraints should also be checked in order to recognize the realizability of the object.

These constraints can be represented by linear inequalities. Suppose, for example, that two faces f_j and f_k share a common edge that has a label "+" and a vertex v_j is on f_k but not on f_j as shown in Figure 10(a). Since the faces form a ridge at the edge, the surface f_j (when extended) should pass through some point between the view point and the vertex v_i. This constraint can be represented by

(9) $$a_j x_i + b_j y_i + z_i + c_j > 0.$$

If the edge has a label "-" as in the case shown in (b), the inequality symbol in (9) is reversed.

Gathering such inequalities, we get a system of linear inequalities

(10) $$B\mathbf{w} > \mathbf{0},$$

where B is a constant matrix and $\mathbf{w}$ is the same unknown vector as in eq. (8). If we construct ineq. (10) in a certain systematic manner, we can prove the following (Sugihara [39, 37]).

Theorem 4. For a type-PAK problem with Assumptions O1, W2, D1, and D2, a polyhedron is realizable if and only if there is a solution of the linear system consisting of eq. (8) and ineq. (10).

By slight modification, the satisfaction of the linear constraints (8) and (10) can be reduced to a problem of judging the existence of a feasible solution to some linear programming problem, for which the solving methods are well established (Dantzig [40]). Thus, the realizability can be checked efficiently. Using this theorem, we can judge that the picture in Figure 7(a) is correct whereas the pictures in Figure 7(b), (c), (d) are incorrect.

7. FLEXIBLE JUDGMENT OF THE REALIZABILITY

7.1. Generic Realizability

Theorem 4 gives a method for strict judgment of the realizability. It is, however, too strict from a practical point of view. Among the four pictures in Figure 7, the picture (a) only is judged correct. This is mathematically right, but we often want to extract the structure of a truncated pyramid from the picture in (b) because digitization errors are inevitable in practical situations such as interpretation of engineering drawings and scene analysis.

As we have observed, some pictures are sensitive to displacements of vertices on the picture plane, while others are not. For example, the picture in Figure 7(a) represents a tetrahedron and this fact remains true even if the vertices are displaced. In contrast, the picture in Figure 7(b) represents a truncated pyramid only when the vertices are drawn in some special position. In order to distinguish the two kinds of pictures, we introduce the following definitions.

A point $\mathbf{p} = (x_1, y_1, ..., x_n, y_n)$ is in *generic position* if $x_1, y_1, ..., x_n, y_n$ are algebraically independent over the rational field. From the definition of algebraic independence, for $\mathbf{p}$ in generic position any polynomial of $x_1, y_1, ..., x_n, y_n$ with rational coefficients is equal to 0 if and only if it is identically equal to 0 when $x_1, y_1, ..., x_n, y_n$ are treated as distinct symbols. In particular, if $\mathbf{p}$ is in generic position, any subdeterminant of the coefficient matrix $A(I, \mathbf{p})$ in eq. (8) vanishes if and only if it is identically equal to 0. Hence, if $\mathbf{p}$ is in generic position, the realizability of $I(\mathbf{p})$ does not depend on $\mathbf{p}$; it depends only on I. An incidence structure I is *generically realizable* if $I(\mathbf{p})$ is realizable for some (and hence any) point $\mathbf{p}$ in generic position.

The generic realizability is characterized by the next theorem, which was conjectured by Sugihara [41], proved by him for some special cases including trihedral polyhedrons and convex polyhedrons (Sugihara [41]), and recently proved completely by Whiteley [42] (see also Sugihara [37]).

Theorem 5. An incidence structure $I = (V, F, R)$ is generically realizable if and only if, for any subset X of F such that $|X| \geq 2$, the inequality

$$(11) \qquad |V(X)| + 3|X| \geq |R(X)| + 4,$$

holds, where $V(X)$ is the set of vertices that are on some faces in X, and $R(X) = R \cap (V \times X)$.

The object represented by the picture in Figure 7(a) has four vertices and three faces (note that we count only visible faces because we consider the picture as a hidden-line-eliminated picture). Since the three faces are all triangular, there are 3×3 = 9 incidence pairs. Putting $X = F$, we get the left hand side of ineq. (11) as $|V| + 3|F| = 13$ and the right hand side as $|R| + 4 = 13$; the inequality is fulfilled. Any two-element subset of F also satisfies it. Thus, from Theorem 5 the incidence structure is generically realizable. For the object shown in Figure 7(b), on the other hand, we get $|V| = 6$, $|F| = 4$, and $|R| = 4\times3 + 3\times1 = 15$ (note that there are three quadrilateral faces and one triangular face). If we put $X = F$, the left hand side of ineq. (11) equals $|V| + 3|F| = 18$ and the right hand side equals $|R| + 4 = 19$; the inequality is not satisfied. Hence, we conclude that the incidence structure is not generically realizable.

Direct application of Theorem 5 is time consuming. The inequality (11) should be checked for almost all subsets of F, and consequently the time required is exponential of the size of the incidence structure. However, efficient algorithms (indeed $O(|R|^2)$ algorithms) are known (Imai [43] and Sugihara [44]).

7.2. Flexible Extraction of Spatial Structures

Using Theorem 5, we can construct a computational mechanism that can extract spatial structures from line drawings flexibly in the sense that the structures can be extracted also from incorrect pictures if the incorrectness is only due to small displacements of vertices (Sugihara [45, 46, 41, 37], Sugihara *et al.* [47]). The outline is the following.

Suppose that a labeled line drawing is given. First, using Theorem 5, we check whether the incidence structure is generically realizable. If so, we apply Theorem 4 directly, because the generic realizableness assures us that the realizability is insensitive to vertex-position errors. If it is not generically realizable, Theorem 4 is useless. Intuitively, this is because the system (8) of linear equations is redundant. If noise is added to the coefficients of a redundant system of equations, it becomes contradictory and admits no solution. This is the reason why a picture whose incidence structure is not generically realizable is sensitive to noise.

In order to circumvent this supersensitivity, we must delete the redundancy in eq. (8). This can be done by removing equations from (8) (in other words, removing incidence pairs from R) one by one until the remaining incidence substructure becomes generically realizable. An $O(|R|^2)$ algorithm is given for this purpose by Imai [48]. Once an generically realizable incidence substructure is found, we apply Theorem 4 to this substructure and judge whether the object is realizable. If realizable, we construct the structure in the space and modify it in order to satisfy the equations that have been deleted. This method works if all the deleted incidence pairs are concerning with the vertices that are on three or less faces in the original incidence structure; see Sugihara [45, 41, 37] for details.

Figure 11 shows an example of flexible extraction of a polyhedral structure by a computer. The picture in (a) is an input line drawing. First it is labeled as shown in (b). next, the incidence structure is recognized as being not generically realizable, and a generically realizable substructure is obtained by removing three incidence pairs, as shown in (c), where the broken lines represent that some incidence pairs are removed so that the object surface may have gaps along these broken lines. A solution is found to the system consisting of eq. (8) and ineq. (10) associated with this substructure, and the three-dimensional structure is constructed as shown in (d).

Another approach to flexible extraction of object structures is to realize interactive systems between men and computers (Lafue [49] and Liardet *et al.* [50]).

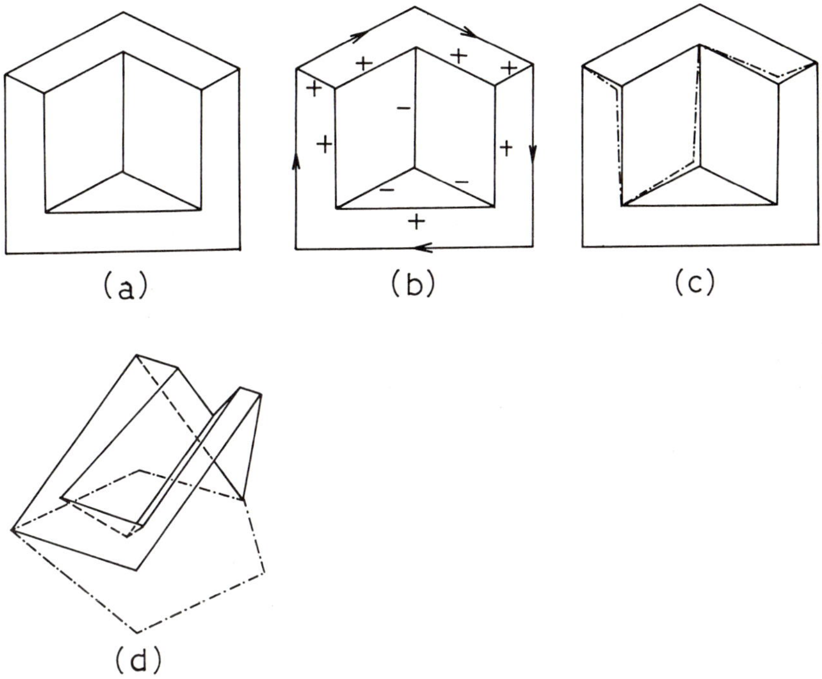

FIGURE 11. Flexible interpretation of a line drawing.

8. REALIZABILITY OF RECTANGULAR OBJECTS

Among the four object worlds given by Assumptions O2, O3, O4, and O5, the rectangular object world is quite different from the others in that the object world is specified by metric properties of the polyhedrons, that is, by angles. As a result, the realizability problem in this object world depends on the view point or the view direction (recall Theorem 1).

We shall review some basic properties of the perspective projection. Suppose that the three-dimensional world is projected onto the xy plane with respect to the view point e. In perspective projection a family of mutually parallel lines in the space are in general mapped to a family of lines having a common point of intersection, say v, called the *vanishing point*. The vanishing point v coincides with the point at which the view line that is parallel to the family of the spatial l ıes penetrates the picture plane, as shown in Figure 12. If there is another family of mutually parallel lines in the space, they are mapped to lines sharing their own vanishing point, say v'. Since the lines ev and ev' are respectively parallel to the two families of spatial lines, the angle formed by the two family of lines equals to the angle θ between the line ev and ev' shown in Figure 12. In particular, if the two families of lines are perpendicular to each other, the angle θ equals $\pi/2$.

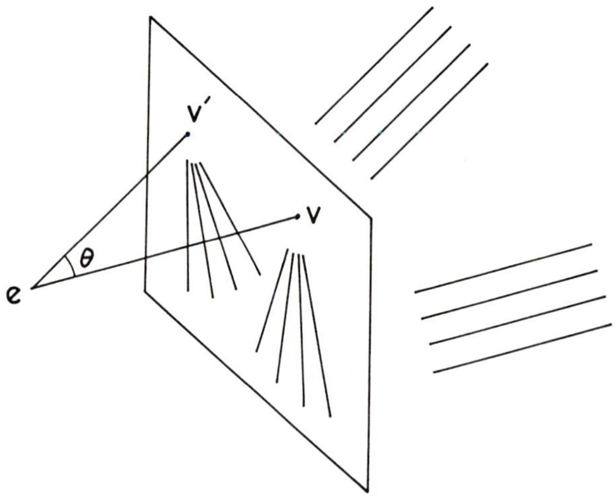

FIGURE 12. View point and vanishing points.

Now we are ready to answer the realizability problems in the rectangular object world. Consider the line drawing shown in Figure 13. Does it represent a rectangular block? To answer the question, we have to search for both a rectangular block and a view point that together give the line drawing. The block has three families of mutually parallel edges, which yield the three vanishing points v, v', and v'' in the picture plane as shown. Let e be the view point. As we observed in Figure 12, the angle between ev and ev' should be $\pi/2$; this implies that the point e should be on the sphere with the diameter vv' (recall that in the two-dimensional space the angle between ev and ev' is $\pi/2$ only when e is on the circle with the diameter vv'). Similarly, both the angle between ev' and ev'' and that between ev and ev'' should be $\pi/2$, and hence e should also be on the sphere with the diameter $v'v''$ and on the sphere with the diameter vv''. Thus, the view point e should be at the common point of intersection of the three spheres. There is only one such point in the front side of the picture plane. This unique view point e is on the line that passes through the point p and is perpendicular to the picture plane, and the distance of e from the picture plane is as indicated by the line segment l in the figure. Thus, the answer to the realizability problem for Figure 13 is affirmative.

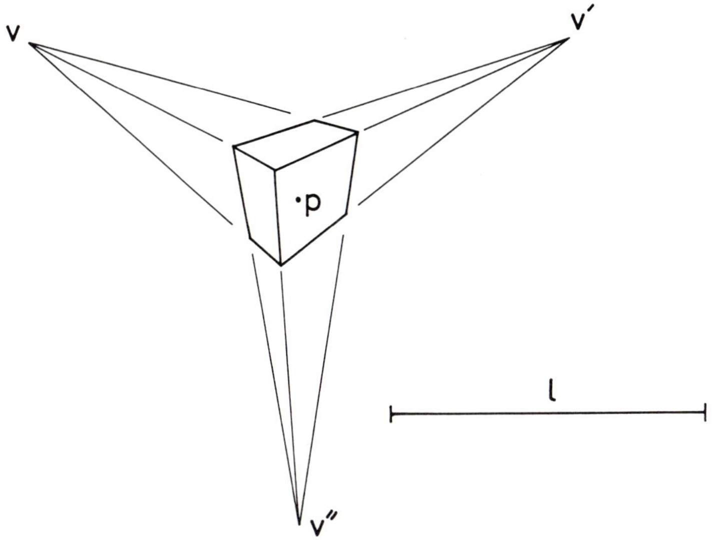

FIGURE 13. Realizable rectangular block.

Next, consider the realizability of a rectangular block from the line drawing given in Figure 14. The three families of mutually parallel edges yield the three vanishing points v, v', and v''. However, the three spheres defined by the diameters vv', $v'v''$, and vv'' do not have a common point of intersection (note that on the picture plane the point v'' is in the circle with the diameter vv'). Thus, a rectangular block is not realizable from this line drawing.

When images of mutually parallel edges are parallel in the picture plane, the corresponding spatial edges are parallel to the picture plane, because otherwise they should have a vanishing point. The picture in Figure 15(a) has two vanishing points v and v', and hence the view point should be on the sphere with the diameter vv'. On the other hand, the third family of lines are mutually parallel in the picture; the direction is indicated by the vector $\mathbf{a}$. The corresponding spatial edges should also be parallel to $\mathbf{a}$, and hence the first two families of spatial edges should be on a plane that is perpendicular to $\mathbf{a}$. This is possible only when vv' is perpendicular to $\mathbf{a}$. This condition is fulfilled by the picture in (a), and hence the block is realizable from the picture. The possible view point is on the circle with the diameter vv' that is on the plane passing vv' perpendicular to the picture plane.

In the picture in (b), on the other hand, the line vv' is not perpendicular to the third family

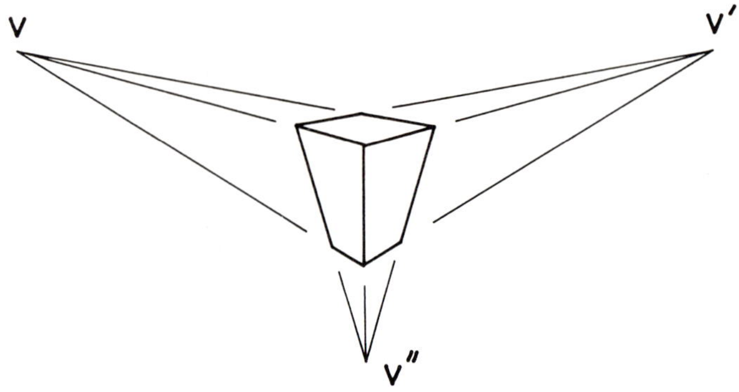

FIGURE 14. Unrealizable rectangular block.

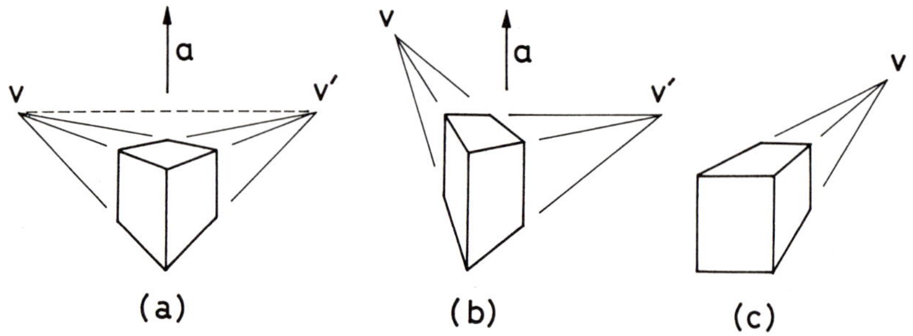

FIGURE 15. Realizability of rectangular blocks from pictures with one or two vanishing points.

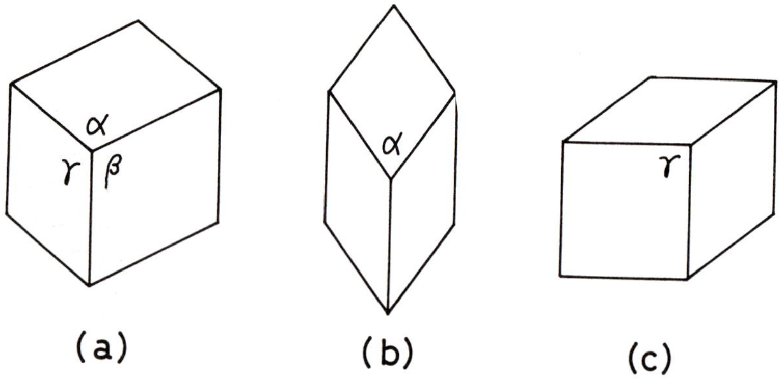

FIGURE 16. Realizability of rectangular blocks in parallel projection.

of parallel edges, and hence a rectangular block is not realizable.

In the picture in (c), two families of lines are mutually parallel in the picture, and hence the left lower face itself is parallel to the picture plane. Consequently, the third family of edges should be perpendicular to the picture plane. This is possible only when the view point is on the line passing through the vanishing point v perpendicular to the picture plane.

In the case of parallel projection, the three families of mutually parallel edges gives three groups of mutually parallel lines in the picture. Consider the pictures in Figure 16. Let α, β, and γ be the three mutual angles formed by the three edges meeting at the central vertex, as shown in the figure.

If we restrict to the orthographic projection, the block is realizable only when the three angles α, β, and γ are all greater than $\pi/2$ [2]. Hence, the block is realizable from the picture in (a), but not realizable from the picture in (b) or (c); in (b) the angle α is less than $\pi/2$, and in (c) the angle γ is equal to $\pi/2$.

If we consider the realizability in parallel projections, the answer is always affirmative. Indeed, any combination of the three angles α, β, and γ can be given by some parallel projection of a rectangular block [2].

The above consideration about blocks can be generalized directly for general rectangular objects, because any rectangular object has exactly three families of mutually parallel edges.

Line drawings of rectangular objects have long been studied in descriptive geometry (for example, Hohenberg [1] and Japan Society for Graphic Science [2]), and hence fundamental properties of them are well established. However, the main motivation is to describe given objects. There are not enough studies from the converse point of view, that is, from the realizability point of view. Nakatani and Kitahashi [13] used vanishing points to check the inconsistency in line drawings of rectangular objects. Kanatani [14] gives a nice summary of mathematical structures of orthographic projection of rectangular objects. Sugihara [51] studies degrees of freedom in realizing objects that are not exactly rectangular but "partially rectangular."

9. CONCLUDING REMARKS

We have briefly reviewed main results in the realizability of polyhedrons from line drawings. The problem itself has many variations, so that the study on this problem admits a wide range of approaches from mathematically strict treatments in terms of graphical geometry and linear algebra to more flexible approaches in terms of heuristics and combinatorics. However, the history is short, and many problems are still left open. For example, a necessary and sufficient condition for polyhedrons to be realizable from a given line drawing is known only for a few variations.

Some of readers may think that it is a powerful check of a line drawing to construct, in terms of descriptive geometry, a second projection of the object seen from another view point. However, it is not true. Note that a line drawing does not convey enough information about the depths. In order to construct the second projection we have to choose and fix the depths of some geometric elements, and use them as the initial constraint to construct the second projection. If we succeed in constructing the second projection, we of course can conclude that the line drawing represents a polyhedron. If we do not succeed, on the other hand, we cannot conclude that the line drawing is incorrect, because another choice of the initial depths may read us to a consistent second projection. Thus, the construction of the second projection gives only a sufficient condition for the realizability.

The realizability is also studied from a psychological point of view. In visual psychology there is a class of pictures called "anomalous pictures" or "pictures of impossible objects" (Penrose and Penrose [52] and Robinson [53]), which give a kind of visual illusion in such a way that, when one see them, one feels that some three-dimensional objects are represented but at the some time one feels something funny. Impossibility of such objects has been studied from mathematical point of view by Cowan [54, 55], Draper [56], Terouanne [57], Sugihara [58], Kupla [59], and Thro [60].

The realizability has an intimate relationship with the rigidity of two-dimensional skeletal structures. We have regarded diagrams as projections of polyhedrons, but we can also interpret the diagrams as skeletal structures composed of rigid rods and rotatable joints. As we saw, a correct picture of a polyhedron has a reciprocal diagram. If we rotate the reciprocal diagram round on the picture plane by $\pi/2$, we get another diagram, where arcs are respectively parallel to the corresponding arcs of the original picture. Hence, arcs in the rotated diagram can be regarded as force vectors acting along the corresponding rods in the skeletal structure. The property that arcs converging to a point in the original picture correspond to arcs forming a closed path in the other implies that the force acting at a joint is in the state of equilibrium. Thus, we see that if a picture is a correct projection of a polyhedron, it can be interpreted as a two-dimensional skeletal structure that admits internal force in equilibrium. This relationship between realizability and force in equilibrium has long been known and used for the graphical calculus of stresses (Maxwell [34, 35], Cremona [36], Crapo [61], and Whiteley [29]). The relationship has been generalized in several directions by Crapo and Whiteley [38], Sugihara [62, 63], and Whiteley [64].

The realizability problem also has other variations. They include realizability from line drawings with additional information such as surface texture and shadows (Kanade [30], Shafer and Kanade [65], Sugihara [46], and Shakunaga and Kaneko [66]), realizability from multiple-view drawings such as three-view drawings (Shapira [67], Idesawa *et al.* [68], Markowsky and Wesley [69], Wesley and Markowsky [70], Preiss [71], and Haralick and Queeney [72]), and realizability from imperfect line drawings (Shapira and Freeman [73] and Shapira [74]).

ACKNOWLEDGMENTS

The author would like to express his thanks to the reviewer, whose comments enable him to improve the manuscript.

REFERENCES

[1] Hohenberg, F., *Konstruktive Geometrie in der Technik*, Springer, Berlin, 1966.

[2] Japan Society for Graphic Science (ed.), *Handbook for Graphic Science* (in Japanese), Morikita-Shuppan, Tokyo, 1980.

[3] Duda, R. O., and Hart, P. E., *Pattern Classification and Scene Analysis*, Wiley, New York, 1973.

[4] Winston, P. H., *Artificial Intelligence*, McGraw-Hill, New York, 1977.

[5] Ballard, D. H., and Brown, C. M., *Computer Vision*, Prentice-Hall, Englewood Criffs, 1982.

[6] Marr, D., *Vision*, Freeman and Company, San Francisco, 1982.

[7] Huffman, D. A., Impossible objects as nonsense sentences, in: Meltzer, B., and Michie, D. (eds.), *Machine Intelligence 6*, Edinburgh University Press, Edinburgh, 1971, pp. 295-323.

[8] Clowes, M. B., On seeing things, *Artificial Intelligence*, vol. 2 (1971), pp. 79-116.

[9] Waltz, D., Understanding line drawings of scenes with shadows, in: Winston, P. H. (ed.), *The Psychology of Computer Vision*, McGraw-Hill, New York, 1975, pp. 19-91.

[10] Sugihara, K., Picture language for skeletal polyhedra, *Computer Graphics and Image Processing*, vol. 8 (1978), pp. 382-405.

[11] Kanade, T., A theory of Origami world, *Artificial Intelligence*, vol. 13 (1980), pp. 279-311.

[12] Sanker, P. V., A vertex coding scheme for interpreting ambiguous trihedral solids, *Computer Graphics and Image Processing*, vol. 6 (1977), pp. 61-89.

[13] Nakatani, H., and Kitahashi, T., Inferring 3-d shape from line drawings using vanishing points, *Proceedings of the First International Conference on Computers and Applications*, Peking, 1984, pp. 683-688.

[14] Kanatani, K., The constraints on images of rectangular polyhedra, *IEEE Transactions on Pattern Analysis and Machine Intelligence*, vol. PAMI-8 (1986), pp. 456-463.

[15] Turner, K. J., Computer perception of curved objects using a television camera, Doctoral dissertation, School of Artificial Intelligence, Edinburgh University, 1974.

[16] Sugihara, K., Automatic construction of junction dictionaries and their exploitation for the analysis of range data, *Proceedings of the 6th International Joint Conference on Artificial Intelligence*, Tokyo, 1979, pp. 859-864.

[17] Montanari, U., Networks of constraints — Fundamental properties and applications to picture processing, *Information Sciences*, vol. 7 (1974), pp. 95-132.

[18] Rosenfeld, A., Hummel, R. A., and Zucker, S. W., Scene labeling by relaxation operations, *IEEE Transactions on Systems, Man and Cybernetics*, vol. SMC-6 (1976), pp. 420-433.

[19] Mackworth, A. K., Consistency in networks of relations, *Artificial Intelligence*, vol. 8 (1977), pp. 99-118.

[20] McGregor, J. J., Relational consistency algorithms and their applications in finding subgraphs and graph isomorphisms, *Information Sciences*, vol. 19 (1979), pp. 229-250.

[21] Haralick, R. M., and Shapiro, L. G., The consistent labeling problem, Part I, *IEEE Transactions on Pattern Analysis and Machine Intelligence*, vol. PAMI-1 (1979), pp.

173-184.

[22] Haralick, R. M., and Shapiro, L. G., The consistent labeling problem, Part II, *ibid.*, vol, PAMI-2 (1980), pp. 193-203.

[23] Nudel, B., Consistent-labeling problems and their algorithms — Expected-complexities and theory-based heuristics, *Artificial Intelligence*, vol. 21 (1983), pp. 135-178.

[24] Kirousis, L. M., and Papadimitriou C. H., The complexity of recognizing polyhedral scenes, *Proceedings of the 26th Annual Symposium on Foundations of Computer Science*, Portland, Oregon, October, 1985, pp. 175-185.

[25] Huffman, D. A., Curvature and creases — A primer on paper, *IEEE Transactions on Computer*, vol. C-25 (1976), pp. 1010-1019.

[26] Huffman, D. A., A duality concept for the analysis of polyhedral scenes, in: Elcock, E. W., and Michie, D. (eds.), *Machine Intelligence 8*, Ellis Horwood, England, 1977, pp. 475-492.

[27] Huffman, D. A., Realizable configurations of lines in pictures of polyhedra, *ibid.*, pp.493-509.

[28] Mackworth, A. K., Interpreting pictures of polyhedral scenes, *Artificial Intelligence*, vol. 4 (1973), pp. 121-137.

[29] Whiteley, W., Realizability of polyhedra, *Structural Topology*, vol. 1 (1979), pp. 46-58.

[30] Kanade, T., Recovery of the three-dimensional shape of an object from a single view, *Artificial Intelligence*, vol. 17 (1981), pp. 409-460.

[31] Draper, S. W., The use of gradient and dual space in line-drawing interpretation, *ibid.*, vol. 17 (1981), pp. 461-508.

[32] Aurenhammer, F., A criterion for the affine equivalence of cell complexes in $\mathbf{R}^d$ and convex polyhedra in $\mathbf{R}^{d+1}$, *Discrete and Computational Geometry* vol. 2 (1987), pp. 49-64.

[33] Ash, P., Bolker, E., Crapo, H., and Whiteley, W., Convex polyhedra, Dirichlet tessellations and spider webs (to appear in: Senechal, M., and Fleck, G. (eds.), *Shaping Space*, Birkhauser, Boston).

[34] Maxwell, J. C., On reciprocal figures and diagrams of forces, *Philosophical Magazine*, ser. 4, vol. 27 (1864), pp. 250-261.

[35] Maxwell, J. C., On reciprocal figures, frames, and diagrams of forces, *Transactions of Royal Society of Edinburgh*, vol. 26 (1870), pp. 1-40.

[36] Cremona, L, *Graphical Statics* (Beare, T. H., trans.), Clarendon Press, London, 1890.

[37] Sugihara, K., *Machine Interpretation of Line Drawings,* MIT Press, Cambridge, 1986.

[38] Crapo, H., and Whiteley, W., Statics of frameworks and motions of panel structures — A projective geometric introduction, *Structural Topology*, vol. 6 (1982), pp. 43-82.

[39] Sugihara, K., A necessary and sufficient condition for a picture to represent a polyhedral scene, *IEEE Transactions on Pattern Analysis and Machine Intelligence*, vol. PAMI-6 (1984), pp. 579-586.

[40] Dantzig, G. B., *Linear Programming and Extensions*, Princeton University Press, Princeton, 1963.

[41] Sugihara, K., An algebraic and combinatorial approach to the analysis of line drawings of polyhedrons, *Discrete Applied Mathematics*, vol. 9 (1984), pp. 77-104.

[42] Whiteley, W., Some matroids on hypergraphs with applications in scene analysis and geometry, preprint, Champlain Regional College (900, Riverside Drive, St. Lambert, Quebec, J4P 3B8, Canada), 1984.

[43] Imai, H., Network-flow algorithms for lower-truncated transversal polymatroids, *Journal of the Operations Research Society of Japan*, vol. 26 (1983), pp. 186-211.

[44] Sugihara, K., Detection of structural inconsistency in systems of equations with degrees of freedom and its applications, *Discrete Applied Mathematics*, vol. 10 (1985), pp. 297-312.

[45] Sugihara, K., Mathematical structures of line drawings of polyhedrons — Toward man-machine communication by means of line drawings, *IEEE Transactions on Pattern Analysis and Machine Intelligence*, vol. PAMI-4 (1982), pp. 458-469.

[46] Sugihara, K., An algebraic approach to shape-from-image problems, *Artificial Intelligence*, vol. 23 (1984), pp. 59-95.

[47] Sugihara, K., Iguchi, H., and Sugie, N., An experimental system for generating polyhedral structures from line drawings, *Journal of Information Processing*, vol. 9 (1986), pp. 135-144.

[48] Imai, H., On combinatorial structures of line drawings of polyhedra, *Discrete Applied Mathematics*, vol. 10 (1985), pp. 79-92.

[49] Lafue, G., A theorem prover for recognizing 2-D representations of 3-D objects, in: Latombe, J.-C. (ed.), *Artificial Intelligence and Pattern Recognition in Computer Aided Design*, North-Holland, Amsterdam, 1978, pp. 391-401.

[50] Liardet, M., Holmes, C., and Rosenthal, D., Input to CAD systems — Two practical examples, *ibid.*, pp. 403-427.

[51] Sugihara, K., Interpretation of axonometric projection of a polyhedron, *Computers and Graphics*, vol. 8 (1984), pp. 391-400.

[52] Penrose, L. S., and Penrose, R., Impossible objects — A special type of visual illusion, *British Journal of Psychology*, vol. 49 (1958), pp. 31-33.

[53] Robinson, J. O., *The Psychology of Visual Illusion*, Hutchinson, London, 1972.

[54] Cowan, T. M., The theory of braids and the analysis of impossible figures, *Journal of Mathematical Psychology*, vol. 11 (1974), pp. 190-212.

[55] Cowan, T. M., Organizing the properties of impossible figures, *Perception*, vol. 6 (1977), pp. 41-56.

[56] Draper, S. W., The Penrose triangle and a family of related figures, *ibid.*, vol. 7 (1978), pp. 283-296.

[57] Terouanne, E., On a class of "impossible" figures — A new language for a new analysis, *Journal of Mathematical Psychology*, vol. 22 (1980), pp. 24-47.

[58] Sugihara, K., Classification of impossible objects, *Perception*, vol. 11 (1982), pp. 65-74.

[59] Kupla, Z., Are impossible figures possible?, *Signal Processing*, vol. 5 (1983), pp. 201-220.

[60] Thro, E. B., Distinguishing two classes of impossible objects, *Perception*, vol. 12 (1983), pp. 733-751.

[61] Crapo, H., Structural rigidity, *Structural Topology*, vol. 1 (1979), pp. 26-45.

[62] Sugihara, K., A unifying approach to descriptive geometry and mechanisms, *Discrete Applied Mathematics*, vol. 5 (1983), pp. 313-328.

[63] Sugihara, K., A correspondence between line drawings of polyhedrons and plane skeletal structures, *The First International Symposium for Science on Form*, Tsukuba, 1985, pp. 273-280.

[64] Whiteley, W., A correspondence between scene analysis and motions of frameworks, *Discrete Applied Mathematics*, vol. 9 (1984), pp. 269-295.

[65] Shafer, S. A., and Kanade, T., Using shadows in finding surface orientations, *Computer Vision, Graphics, and Image Processing*, vol. 22 (1983), pp. 145-176.

[66] Shakunaga, T., and Kaneko, H., Perspective angle transform and its application to 3-D configuration recovery, *Proceedings of IEEE Conference on Computer Vision and Pattern Recognition*, 1986, pp. 594-601.

[67] Shapira, R., A technique for the reconstruction of a straight-edge, wire-frame object from two or more central projections, *Computer Graphics and Image Processing,*, vol. 3 (1974), pp. 318-326.

[68] Idesawa, M., Soma, T. Goto, E., and Shibata, S., Automatic input of line drawing and generation of solid figure from three-view data, *Proceedings of International Computer Symposium*, 1975, vol. II, pp. 304-311.

[69] Markowsky, G., and Wesley, M. A., Fleshing out wire frames, *IBM Journal of Research and Development*, vol. 24 (1980), pp. 582-597.

[70] Wesley, M. A., and Markowsky, G., Fleshing out projections, *ibid.*, vol. 25 (1981), pp. 934-954.

[71] Preiss, K., Algorithms for automatic conversion of a 3-view drawing of a plane-faced part to the 3-D representation, *Computers in Industry*, vol. 2 (1981), pp. 133-139.

[72] Haralick, R. M., and Queeney, D., Understanding engineering drawings, *Computer Graphics and Image Processing*, vol. 20 (1982), pp. 244-258.

[73] Shapira, R., and Freeman, H., The cyclic order property of vertices as an aid in scene analysis, *Communications of ACM*, vol. 22 (1979), pp. 368-375.

[74] Shapira, R., The use of objects' faces in interpreting line drawings, *IEEE Transactions on Pattern Analysis and Machine Intelligence*, vol. PAMI-6 (1984), pp. 789-794.

Computational Morphology
G.T. Toussaint (editor)

VORONOI AND RELATED NEIGHBORS ON DIGITIZED TWO-DIMENSIONAL SPACE WITH APPLICATIONS TO TEXTURE ANALYSIS

by Jun-ichiro TORIWAKI, and Shigeki YOKOI

Department of Information Engineering, Faculty of Engineering, Nagoya University

Furo-cho, Chikusa-ku, Nagoya, 464-01 JAPAN

In this article, we present neighbor relations among figures on a 2-D digitized picture. First those relations (the Voronoi neighbor, the relative neighbor, and the Gabriel neighbor) are extended to a set of connected components (arbitrary shapes of figures). Second, several procedures to obtain those neighbor relations from a given binary image are described with estimation of the amount of computation required. Finally applications of the adjacency graphs induced from the above neighbor relations to texture analysis are described with the results of experiments.

1. Introduction

Computational geometry which mainly treats geometrical properties of a finite point set has attracted increasing attention in the field of image processing [1], Especially the Voronoi neighbor, the relative neighbor and the Gabriel neighbor are expected to be useful as a tool for extracting features of a set of figures scattering in a two-dimensional picture. Usefulness of concepts and methodologies of the computational geometry as a shape factor in image processing is suggested by the term computational morphology appeared in Ref. [2]. Typical examples are applications of the Dirichlet tessellation (DT), the Voronoi diagram (VD), the Gabriel neighbor (GN) and the relative neighbor (RN). Assuming that a finite point set in a continuous space is given the DT or the VD provides a method for dividing the continuous space into mutually disjoint subspaces according to the nearest neighbor rule, and the RN and the GN introduce neighborhood relations into a point set. Furthermore, in order to analyze geometrical properties of a finite point set, arrangement of points in the two-dimensional plane is transformed into a graph by considering each point as a node and connecting two adjacent nodes by a branch. Typical adjacency relations used are the Voronoi neighbor (VN), the Gabriel neighbor (GN) and the relative neighbor (RN) [3]. However, past works on these neighboring relations were of limited significance for digital image processing because:

(1) properties inherent to the processing of digitized pictures have not been considered enough;

(2) no effective algorithm has been shown for deriving the above neighbor relations on a 2-D digitized picture from a given input picture; and

(3) those neighbors have been defined only for a set of isolated points and very limited type of figures such as circles and line segments.

In this article, we present neighbor relations among figures on a 2-D digitized picture. First those relations (the Voronoi neighbor, the relative neighbor, and the Gabriel neighbor) are extended to a set of connected components (arbitrary shapes of figures). Second, several procedures to obtain those neighbor relations from a given binary image are described with estimation of the amount of computation required. Finally applications of the adjacency graphs induced from the above neighbor relations to texture analysis are described with the results of experiments.

Before proceeding to the details of research results we will point out major difference between neighbor relations on the continuous space and those on the digitized 2-D space.

(1) Specific distance measures such as the 4-neighbor distance and the 8-neighbor one are employed for the digitized picture processing instead of the Euclidean distance in the continuous space.

(2) Algorithms which are easily implemented and work well for real digital pictures are required to calculate the neighboring relations effectively instead of procedures for theoretical estimation of the amount of computation.

(3) Neighboring relations should be defined for a set of connected components (figures) as well as for a set of isolated pixels.

(4) Estimation of the amount of computation given in the form of O(***) which is very popular for neighbor relations in the continuous space is not critically significant for the digitized space, because in digital image processing always a finite size of images including finite number of pixels are processed instead of the infinite continuous space. Since the picture size takes very limited range of values (typically $2^n x 2^n$ pixels, n=(8~12)) the estimation of the computation time in the form of O(picture size) is of little importance.

2. Definition of Modified Digital Voronoi Diagram

2.1 Basic definitions

In this article, we consider only digitized pictures sampled at points on the square lattice. Let (i, j) denote a picture element (pixel) located in the ith row and the jth column, and $F=\{f_{ij}\}$ denote a digitized picture in which a value of density at a pixel (i, j) is given by f_{ij}. We assume that $f_{ij}=1$, or, 0 (a binary picture). A t-neighbor distance (t=4 or 8) between two pixels x=(i, j) and y=(p, q) is defined as follow:

$$d_4(x,y) \equiv | i - p | + | j - q | ,$$
$$d_8(x,y) \equiv \max\{| i - p | , | j - q |\} \qquad (1)$$

Only the above two distance measures are used throughout the paper. Pixels with values 0 and 1 are called 0-pixels and 1-pixels, respectively. A set of all 1-pixels is called a <u>figure</u> and its complement (a set of all 0-pixels) is called the <u>background</u>. For a pixel x=(i, j), a set of pixels y such that $d_t(x,y)=1$ is called the t-neighborhood of the pixel x and denoted by $N_t(x)$ or $N_t((i,j))$ (t=4 or 8).

Denoting by C a connected component of 1-elements, the distance between the component C and a pixel x, $d_t(x,C)$ is defined by

$$d_t(x,C)=\min_{y\in C} d_t(x,y). \qquad (2)$$

The distance between two components C1 and C2 is defined by

$$d_t(C1,C2)=\min_{x,y}\{\, d_t(x,y);\; x \in C1,\; y \in C2 \,\} \qquad (3)$$

Using these basic definitions, we define the Modified Digital Dirichlet Tessellation (MDDT) and the Modified Digital Voronoi Diagram (MDVD) as follow.

[Definition 1] (Tile) Assuming a picture F contains n connected components C1,C2, ···,Cn, a set of pixels Tr defined below is called a tile of the connected component Cr.

$$Tr \equiv \{x;\; d_t(x,Cr) < d_t(x,Ck),\; \forall k \neq r\} \qquad (4)$$

A set of all tiles Tr($1 \leq r \leq n$), and division of a picture F into such tiles are called the MDDT of the picture F and denoted by T(F).

The tile Tr is a set of pixels to which the connected component Cr is closest of all connected components. Obviously, Cr $\subseteq$ Tr. The set of all such tiles is the MDDT. The MDVD is defined as the set of all pixels which are located on borders of tiles and all pixels which belong to no tile (or pixels which have more than one closest components to them). The following gives more strict definition.

[Definition 2] (MDVD) Under the same assumption as Definition 1, a set of pixels V(F) defined below is called the MDVD of the picture F.

$$V(F) \equiv \{x;\; x \notin (\bigcup_{k=1}^{n} T_k) \cup [\bigcup_{k=1}^{n} \{x;\; x \in T_k,\; \exists y \in T_r,\; r \neq k,\; d_t(x,y)=1\}] \qquad (5)$$

The righthand side of the equation includes two terms combined by the logical sum. The first term represents the set of pixels which belong to no tile. The second means the set of all pixels which are adjacent to at least one tile different from their own tile.

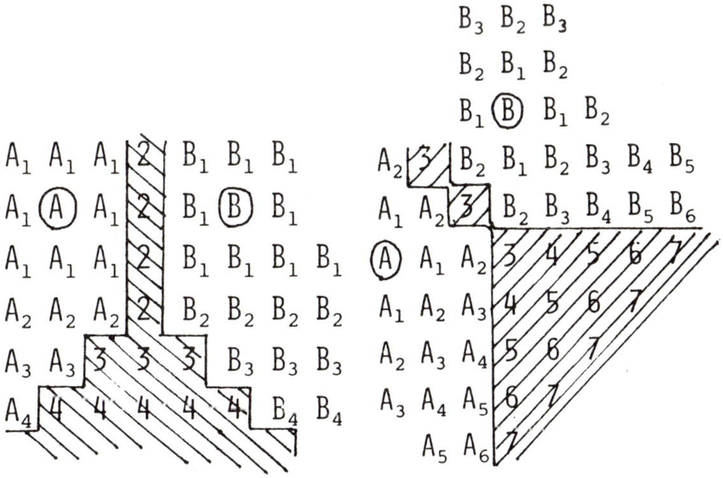

Fig. 1. Sets of pixels which are of equidistance from two pixels-in the case of d_8 (left) and d_4 (right).

Unfortunately we found a serious difficulty in this definition that the MDVD which should be of linear form may become an infinite half-plane [4.5]. This is a characteristic feature of the t-neighborhood distance measure (Fig. 1). Two alternatives are derived by employing

(1) the definition by a product-sum form of pixel sets, and

(2) the structural form of the definition by a recursive procedure [4].

The former is given in the Appendix and the latter is shown below.

[Definition 3] Denote a label picture in which different integers are given to different connected components to identify each component in the input picture F by $A(0)=\{a_{ij}(0)\}$ (where $a_{ij}(0)$ is 0, if (i,j) is a 0-pixel and an integer identifying connected component to which (i,j) belongs if (i,j) is a 1-pixel). Generate a picture sequence $A(k)=\{a_{ij}(k)\}$, k=1,2, ··· from the picture A(0) by the following procedure.

$$a_{ij}(k) = \begin{cases} m, & \text{if } (a_{ij}(k-1)=0) \wedge [\text{in } A(k-1), \text{ at least one positive label exists in the } t_1\text{-neighborhood of } (i,j) \text{ and only one kind of a label (denoted by } m) \text{ exists, where } t_1 \text{ represents the type of neighborhood}] \\ -1, & \text{if } (a_{ij}(k-1)=0) \wedge [(\text{in } A(k-1), \text{more than one kind of labels exist in the } t_1\text{-neighborhood of } (i,j)) \vee (\text{in } A(k-1), \text{ all pixels in the } t_1\text{-neighborhood of } (i,j) \text{ have the value } -1)] \\ a_{ij}(k-1), & \text{if otherwise.} \end{cases} \quad (6)$$

The iteration terminates if A(N) = A(N+1) for a certain positive value N.

The resultant picture A(N) (denote by A) is called the "<u>MDDT</u>" of the picture F.

Each region filled with one label is called a tile of a connected component having the same label.

[<u>Definition 4</u>] A set of pixels V'(F) obtained from a picture F by the following equation is called the <u>MDVD</u> of F.

$$V'(F) = \{(i,j): [a_{ij}=-1] \vee [\exists (p,q)\in N_{t_2}((i,j)), a_{ij}>0, a_{pq}>0, a_{pq}\neq a_{ij} \text{ and } f_{ij}=0]\} \quad (7)$$

where t_2 represents the type of neighborhood used here.

The neighborhood t_2 may be different from the neighborhood t_1 used in Definition 3. The condition $f_{ij}=0$ is needed to accept the MDVD only in the background of the input picture.

The procedure of Definition 3 may be explained in analogy with wave propagation on the picture plane. That is, let the label of each connected component start simultaneously at the corresponding connected component and propagate in a constant velocity in all directions. If more than one different labels arrives at a pixel at the same time, the pixel belongs to no tile and is given the label −1. The difficulty presented just after Definition 2 is avoided by forbidding the wave pass through the pixel having the label −1.

3. Algorithm to Obtain the MDVD

Definition 3 of the MDDT suggests a parallel algorithm to obtain it. In fact, starting from a given label picture, we could replace each 0-pixel P0 in the following way.

[Procedure 1] (i) If only one kind of label exists in its neighborhood (4- or 8-neighborhood) give that label to P0.
(ii) If more than one different labels exists in its neighborhood, give the value −1 to P0.
(iii) Otherwise, keep the value 0 unchanged.

By repeating this procedure until no pixel remains to be changed, we obtain the MDDT. The procedure to calculate the MDVD from the MDDT will be self-evident from Definition 4.

Derivation of the sequential algorithm is not so straightforward. Consideration of similarity of the above definition to the distance transformation [6] leads to the following procedure.

Let us define four scanning modes as are shown in Fig. 2. Let $\hat{F}=\{\hat{f}_{ij}\}$ denote a picture derived by inverting all pixel values of an input binary picture F, that is, $\hat{F}=\{1-f_{ij}\}$, where $\hat{F}$ is called an inverted picture of F.

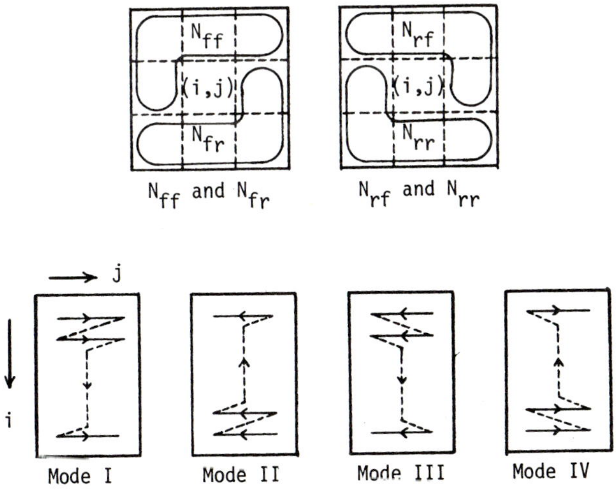

Fig.2 Illustrations for Algorithm 1.

[Algorithm 1] Let F denote an input binary picture.
(Step I) (1.1) Calculate the distance transformation G (t_1- neighborhood type, where t_1=4, 8) [6] of the inverted picture $\bar{F}$.
(1.2) Generate a picture sequence
{A(4k+1), A(4k+2), A(4k+3), A(4k+4); k=0, 1, 2, ···}
from the label picture A(0) of the input picture F and the distance picture G, by repeating the procedures (i) through (iv) given below for k=0, 1, 2, ···.
(i) Obtain A(4k+1) from A(4k) according to the following equation. (Two sets Sf and Sr used there are defined as in Table 1).

$$a_{ij}(4k+1) = \begin{cases} m, & \text{if all pixels of A(4k+1) having positive values in the Sf and all pixels of A(4k) having positive values in the Sr have the identical value m (Note here that values of A(4k+1) have been determined in the Sf),} \\ -1, & \text{if at least two different positive values exist in the union of Sf in the A(4k+1) and the Sr in the A(4k),} \\ a_{ij}(4k), & \text{if otherwise.} \end{cases} \quad (8)$$

(ii) Determine A(4k+2) from A(4k+1) in the same way.
(iii) Determine A(4k+3) from A(4k+2) in the same way.
(iv) Determine A(4k+4) from A(4k+3) in the same way.
Use the raster scanning mode shown in Table 1 and Fig. 2 for each of (i)-(iv).
(Stopping rule) If A(N)=A(N+1) for a certain value of k(for example, k=N), define as A≡A(N). The value -1 is given to a pixel x in A if all of t_1-neighborhood of x have the value -1.
(Step II) The set V'(F) is obtained from the above picture A by the same procedure as shown in Definition 4.

The picture A and the set V'(F) give the MDDT and the MDVD of F, respectively.

The parallel algorithm and the sequential one are regarded as exactly equivalent in principle, because both are controlled by the well-known distance transformation (of the parallel type for the former and the sequential type for the latter) [6]. Furthermore both gave exactly the same results in all of our experiments. Nevertheless no theoretical proof has been given concerning equivalence of these two algorithms.

4. Neighboring Relations

The most important significance of the MDVD from the viewpoint of image analysis and recognition is introduction of the neighboring relation into a set of connected components on a digitized picture plane.

Before proceeding to the neighboring relations on a digitized picture, we will summarize those in the continuous space. Suppose a set of points $\underline{P}$={P1, P2, ···, Pn} is considered on the continuous two-dimensional space. Two points P_i and P_j ($P_i \in \underline{P}$, $P_j \in \underline{P}$) are said to be
(1) the Voronoi neighbor (VN) if there exists a circle passing both P_i and P_j such that no other point of $\underline{P}$ is included inside

it;

(2) the Gabriel neighbor (GN) if no point of $\underline{P}$ exists inside the circle with the diameter $\overline{P_iP_j}$; and,

(3) the relative neighbor (RN) if any point of $\underline{P}$ does not exist in the area common to the two circles with the radius $\overline{P_iP_j}$ and the centers at P_i and P_j, respectively. (Fig. 3)

Next let us extend these concepts of neighboring relation to a digitized picture plane.

4.1 Modified Digital Voronoi Neighbor

Assuming that a picture F contains a set of n connected components $\underline{C}=\{C1, C2, \cdots, Cn\}$, the modified digital Voronoi neighbor (MDVN) is defined basing upon the modified digital Voronoi diagram (MDVD) as follows.

[Definition 5] Two components Ci and Cj are said to be the modified digital Voronoi neighbor (MDVN) each other if any of the following (i) and (ii) holds:

(i) $\exists x \in Ti,\ \exists y \in Tj,\ x \in N_8(y)$
(ii) $\exists x \in Ti,\ \exists y \in Tj,\ \exists z \in B,\ x \in N_8(Z),\ y \in N_8(Z)$

where Ti denotes the tile of the component Ci and B denotes a set of pixels given the value −1 in the MDDT.

Intuitively, two components Ci and Cj in a picture F are MDVN if the corresponding tiles Ti and Tj in the MDVD share parts of their borders. An example is shown in Fig. 4.

4.2 Modified Digital Gabriel Neighbor

In extending the Gabriel neighbor and the relative neighbor, we will use a square defined as follows instead of circles in Fig. 3.

[Definition 6] Given an arbitrary point g and a positive constant r, a set of pixels St(g, r) defined by Equation (9) is called a square with the size r and the center at g;

$$St(g, r)=\{x;\ d_t(x, g) < r\},\quad t=4, 8 \tag{9}$$

Now the digital Gabriel neighbor is defined as follows.

[Definition 7] Given a set of isolated 1-pixels $\underline{P}=\{P1, P2, \cdots, Pn\}$, two pixels Pi and Pj are said to be the digital Gabriel neighbor (DGN) each other if a square $St(c_{ij},\ d_t(c_{ij}, P_i))$ with the size $d_t(c_{ij}, P_i)$ (a forbidden region) does not include any other pixel of P inside it, where c_{ij} is the midpoint of the segment $\overline{P_iP_j}$ (c_{ij} need not be a grid point).

Furthermore the DGN is extended to a set of connected components in the definition below.

[Definition 8] Given a set of connected components $\underline{C}=\{C1, C2, \cdots, Cn\}$, consider two components of $\underline{C}$, C_i and C_j. Let x_k and y_k ($x_k \in C_i$, $y_k \in C_j$) ($k=1, 2, \cdots, n$) denote 1-pixel pairs such that $d_t(x_k, y_k)=d_t(C_i, C_j)$, and let Gk denote a forbidden region for x_k and y_k (Fig. 5). Then, two components C_i and C_j are said to be the modified digital Gabriel neighbor (MDGN) each other if

$$\exists k,\ \forall m(\neq i, j),\ Gk \cap C_m = \phi \text{ (empty set).} \tag{10}$$

Table 1 Explanation of notations in Algorithm 1

$S_f = \{(p,q);\ g_{ij} = g_{pq} + 1,\ (p,q) \in S'_f\}$

$S_r = \{(p,q);\ g_{ij} = g_{pq} + 1,\ (p,q) \in S'_r\}$

$G = \{g_{ij}\}$ = distance transformation (t_1-neighborhood type) of the inverted picture $\hat{F}$

No. of steps	Scanning mode*	S'_f	S'_r
(i)	1	$N_{ff} \cap N_{t_1}((i,j))$	$N_{fr} \cap N_{t_1}((i,j))$
(ii)	2	$N_{fr} \cap N_{t_1}((i,j))$	$N_{ff} \cap N_{t_1}((i,j))$
(iii)	3	$N_{rf} \cap N_{t_1}((i,j))$	$N_{rr} \cap N_{t_1}((i,j))$
(iv)	4	$N_{rr} \cap N_{t_1}((i,j))$	$N_{rf} \cap N_{t_1}((i,j))$

* See Fig. 2

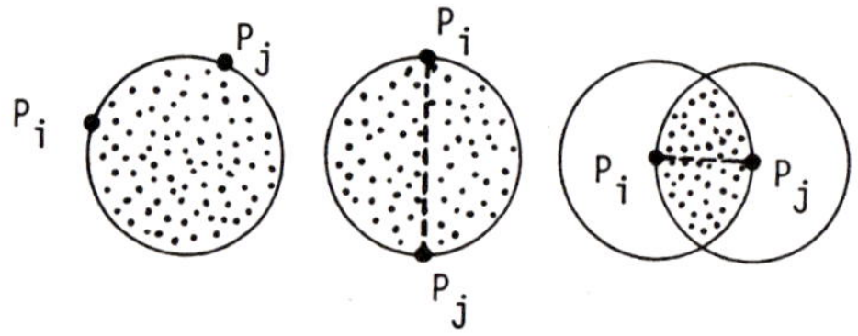

Fig.3 Forbidden regions for Voronoi(left), Gabriel(center), and relative(right) neighbors

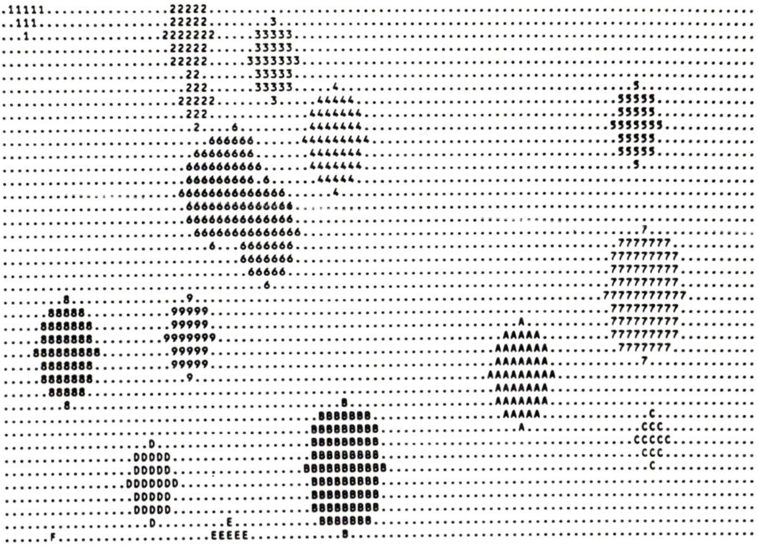

(a)

(b)

Fig. 4. Example of the MDVD (part of a larger picture)
(a) Input image (b) Tiles or MDDT (c) MDVD

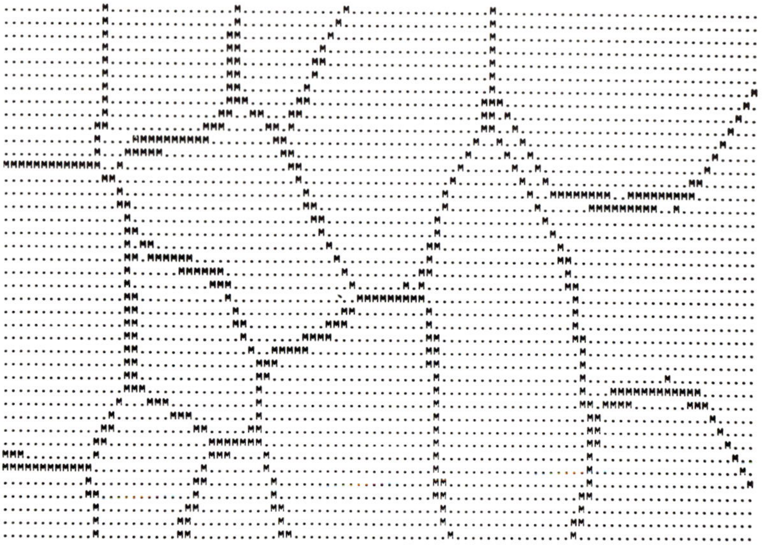

Fig. 4 (continued) (c)

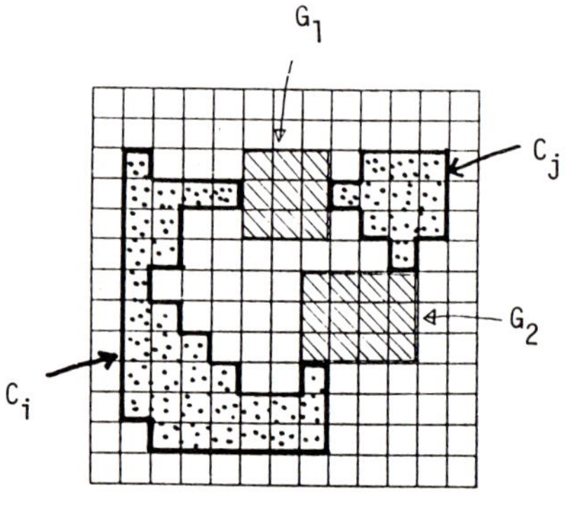

Fig.5 Forbidden region for the MDGN

Intuitively speaking, two components C_i and C_j are MDGN if there exists at least one pair of 1-pixels $x \in C_i$ and $y \in C_j$ of which forbidden region includes no 1-pixels of the component other than C_i and C_j. Two procedures are available to determine the DGN relations among all pixels of a set $\underline{P}$ by iterating the following Procedure (2) or (3) for all pairs of pixels in $\underline{P}=\{P1, P2, \cdots, Pn\}$.

[<u>Procedure 2</u>] Calculate the midpoint c_{ij} for $\overline{P_iP_j}$, and test whether the inequality $d_t(c_{ij}, P_i) \leqq d_t(c_{ij}, P_k)$ holds or not for all P_k's except P_i in $\underline{P}$.

[<u>Procedure 3</u>] Test whether a 1-pixel exists or not in the forbidden region for P_i and P_j.

In the above Definition 8, if there are more than one 1-pixel pairs x_k and y_k such that $d_t(x_k, y_k)=d_t(C_i, C_j)$, $k=1, 2, \cdots, K$, equation (10) must be satisfied by all forbidden regions Gk, $k=1, 2, \cdots, K$, in order that C_i and C_j are MDGN (Fig. 5).

4.3 Modified Digital Relative Neighbor

The digital relative neighbor (DRN) and the modified digital relative neighbor (MDRN) are also derived by extending the relative neighbor by the way similar to the DGN and the MDGN.

[<u>Definition 9</u>] Given a set of isolated 1-pixels $\underline{P}=\{P1, P2, \cdots, Pn\}$, two pixels P_i and P_j are said to be the digital relative neighbor (DRN) each other if there exists no 1-pixel in the area common to two squares $St(P_i, d_t(P_i, P_j))$ and $St(P_j, d_t(P_i, P_j))$ (forbidden region for P_i and P_j).

The MDRN is defined by quite the same way as that of MDGN. The same procedures as those for DGN and MDGN are also applied to the DRN and MDRN.

5. Properties of Neighborhood Relations

5.1 Inclusive relations

[<u>Property 1</u>] DGN $\supseteq$ DRN and MDGN $\supseteq$ MDRN in the sense that
(1) for a set of isolated 1-pixels $\underline{P}=\{P1, P2, \cdots, Pn\}$, if P_i and P_j are DRN, then they are DGN;
(2) for a set of connected components $\underline{C}=\{C1, C2, \cdots, Cn\}$, if C_i and C_j are MDRN, they are also MDGN.

Inverse of Property 1 is not true in general. There exists no inclusive relation among MDVN and MDGN (or MDRN), because configuration of tiles on a digitized picture plane is very complicated and is not predictable theoretically.

5.2 Computation time

Calculation of the Voronoi diagram for the set of points in the continuous space requires the amount of computation O(nlogn) where n is the number of points in the given point set. This estimation is not valid for the MDVD.

The number of times of iterations in the procedure of Definition 3 (or the parallel procedure presented in 3) is about half the maximum width of the background in an input picture. On the other hand, the number of times of iterations in Algorithm 1 cannot be estimated theoretically. In our experiments, the algorithm always terminated after four iterations if the 4-neighbor distance is adopted, while if the 8-neighbor distance is used, more than four times of iterations were required. Thus the number of times of iterations may not be related explicitly to the number of connected components. Generally speaking, the parallel algorithm (Definition 3 and Procedure 1) requires more iterations than the sequential one in Algorithm 1.

In the parallel algorithm, however, computation time for one iteration rapidly decreases with the number of times of iterations because of rapid decrease in the number of pixels whose labels are undefined. In the sequential algorithm, on the other hand, every iteration takes almost the same computation time because in each iteration one must check at all pixels whether or not their labels have been fixed. Therefore, the sequential algorithm is advantageous if the 4-neighbor distance is employed and in particular, the size of an input picture is very large. For other cases we cannot determine absolutely which of the sequential and the parallel type is more advantageous.

The algorithms developed for the continuous space do not seem to be effective in the case of a connected component set because they must consider all possible pairs of points (pixels) which belong to different components each other.

Concerning the modified digital Gabriel neighbor (MDGN), two procedures we presented in 4.2 are considered. In Procedure 2, the computation time is $O(n^3)$, independent of the picture size. In Procedure 3, the computation time increases more gradually with n, because the size of a forbidden region decreases on the average as n increases. But it increases with the picture size.
Table 2 shows several results about the amount of computation.

6. Applications - Texture Analysis Using Adjacency Graphs

6.1 Image features extracted from the adjacency graph

The most important contribution of the neighboring relations, the MDVD, the MDRN and the MDGN to image processing is that they provide tools for introducing adjacency relations to a set of connected components scattered in a digitized picture, and those for deriving an adjacency graph description of randomly scattered figures.

The adjacency graph may be given attributes to its nodes and edges, where nodes correspond to connected components and edges represent adjacent component pairs (Fig. 6). For example, various shape features of individual components are used as nodes attributes, and the distance between adjacent components is considered as an edge feature. Furthermore, features of the adjacency graph are extracted to classify, analyze and recognize the original image. For example, cooccurrence matrices [7] of those features obtained from connected components or tiles neighboring in the above sense are effectively used for quantitative measurement and classification of such random components patterns (Fig. 7). Several examples of such features are shown in Table 3.

Following sections show possible applications of the neighboring relations to texture analysis and recognition. Textures are complex patterns recognized in a two-dimensional image, usually

Table 2 Computational complexity of the procedure to obtain neighbor relations (worst case estimation)

	Procedure 2	Procedure 3
DVN	$O(n^3)$	$O(Ln^2)$
DGN	$O(n^3)$	$O(Ln^2)$
DRN	$O(n^3)$	$O(Ln^2)$
MDVN	$O(m^3n^3)$	$O(Lm^2n^2)$
MDGN	$O(m^3n^3)$	$O(Lm^2n^2)$
MDRN	$O(m^3n^3)$	$O(Lm^2n^2)$

n : Number of isolated pixels or connected components
m : Maximum number of pixels for a connected component
L : Size of input picture (number of pixels)

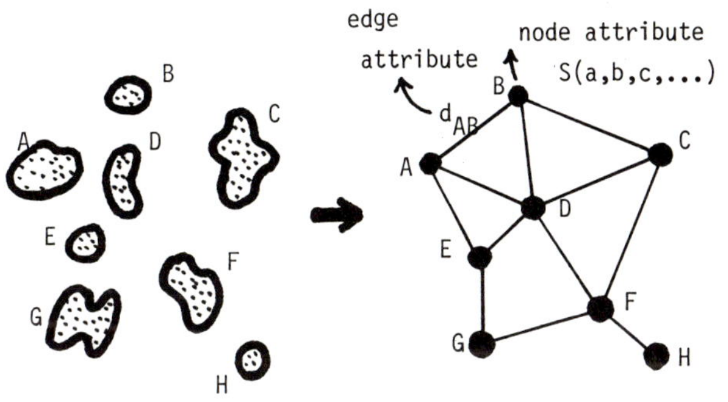

Fig. 6 Adjacency graph induced from neighbor relations

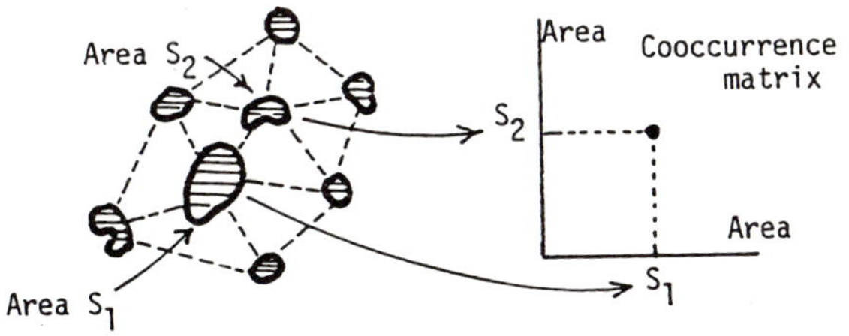

Fig. 7. Adjacency graph and cooccurrence matrix

composed of subpatterns that have various characteristic features such as brightness, colors, shapes, sizes, etc. In most cases, perception of visual textures in human eye is performed by recognizing local areas in which the above features look like homogeneous. Machine recognition of texture patterns contains problems of texture discrimination and texture edge detection. The adjacency graphs basing upon the neighboring relations will be applicable to both of them as are presented in the following sections.

Table 3 Shape features of tiles and graphs

<u>tile feature</u> (=node attribute)

1 Area A
2 Peripheral length L
3 Elongatedness L / A
4 Center of gravity (CG) , or shift of CG of a tile from the original figure (ACG)
5 Height (H) and width (W) of a circumscribed rectangle
6 Ratio of area RS1 =A/H W
7 Ratio of area RS2 =A/ (area of the original figure)

<u>tile-pair feature</u> (=edge attribute)
1 Distance between adjacent components
2 Relative direction of the adjacent components

<u>graph feature</u>
1 Number of neighboring components (tiles)
(=number of edges connected to a node)
2 Length of edges (=distance to the neighboring components)
3 Statistics of shape or graph features at each node
(average, variance etc.)

6.2 Texture region extraction

In the case of MDVN, regions with specific textural features can be extracted by selecting tiles with suitable properties. One such procedure is as follows.

[<u>Procedure 4</u>] (1) Obtain the MDVD and the MDVN-graph from an input binary image.

(2) For each tile in the MDVD, calculate the mean and the variance of the suitable tile features (e.g. area) over the neighboring tiles.

(3) Extract connected components corresponding to the tile for which the mean etc. obtained in (2) is smaller (or larger) than a prespecified threshold.

Above procedures were applied to several natural textures (Fig. 8). Regions having textures different from background were extracted reasonably well. In Fig. 9, very small irregularities in a point pattern are enhanced remarkably by marking "deformed" tiles.

6.3 Texture edge detection

A texture edge (or a border between regions of different textures) may be detected by considering the fact that the edge length in the neighbor graph will be longer in the subregion of a

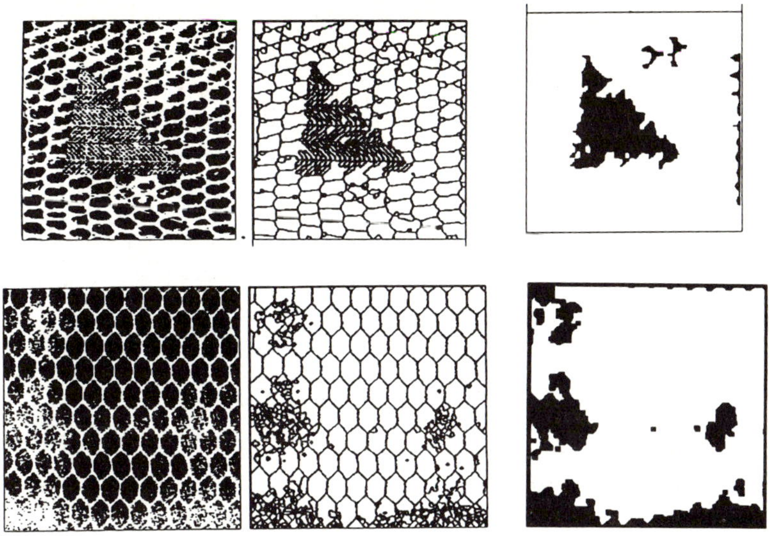

Fig. 8 Texture region extraction
For each row, left: original picture, center: MDVD,
right: extracted texture region

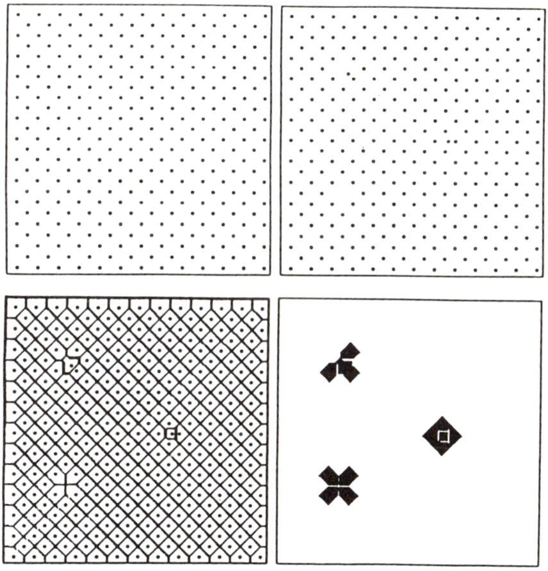

Fig. 9. Deformation detection
top left : regular pattern, top right : pattern with dislocations
bottom left : MDVD of the above pattern,
bottom right : "deformed" tiles

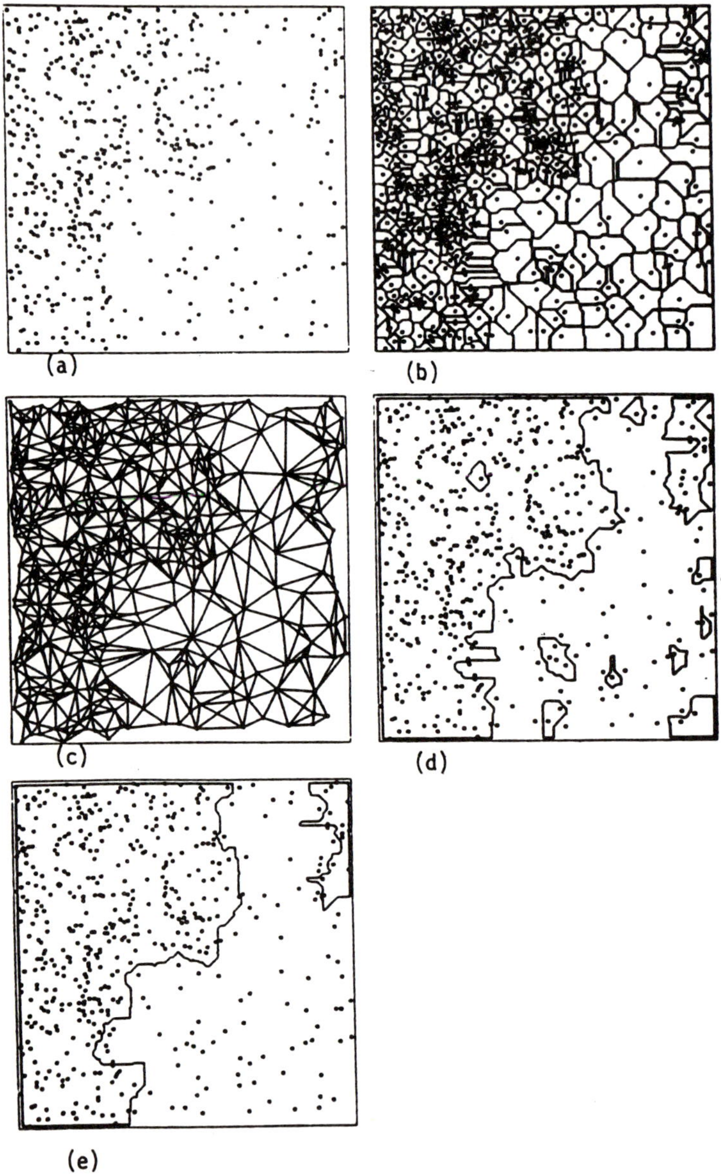

Fig . 10 Texture edge detection
(a) input pattern (b)MDVD (c) adjacency graph
(d) texture edge (First order neighbor was considered)
(e) texture edge (Second order neighbor was considered)

picture where connected components exist sparasely than in the region where they are densely distributed. An important merit of the MDVD is that the texture edge is determined by extracting a suitable set of tiles and following the border of the extracted tile set. The procedure below gives an example.

[Procedure 5] (Processing of an isolated pixel set)
(1) Derive the MDVD-graph from an input picture.
(2) At each node, calculate the average of edge lengths over all edges connected to it. First and second order neighbors are considered here.
(3) Extract nodes for which the above averages larger than a threshold T, and eliminate the corresponding tiles.
(4) Extract borders of a set of remaining tile. Fusion may be applied before border extraction, if necessary.

The second and the third step of the procedure is similar to the method used for clustering of a point set in [10]. The significant point here is to perform segmentation of picture plain by combining the MDVD and clustering of a point set.

Some results are shown in Fig. 10. Smoother border is extracted by considering higher order neighbors.

6.4 Texture discrimination by the cooccurrence matrix

In order to classify textural patterns we need features to characterize whole of a texture pattern instead of subpatterns or tiles containd in it. As one of such features we can utilize a new type of cooccurrence matrix, that is, a two-dimensional histrogram of feature values which are taken by pairs of connected components neighboring each other according to the neighbor relations discussed here (Fig. 7). The following is an example of possible procedures.

[Procedure 6] (1) Derive a suitable neighbor graph from an input picture and determine neighboring component pairs from the graph.
(2) Calculate the coocurrence matrix of features of component pairs.
(3) Extract necessary features from the matrix.

Experimental results are shown in Fig. 11 and 12. The area of tile was used as a tile feature. To classify textures we need to calculate features of cooccurrence matrix. From typical features shown in Ref. [7], we employed the contrast and the energy, which are defined as follows.

Contrast : $$C = \sum_{k=0}^{m-1} k^2 \left\{ \sum_{f_i=1}^{m} \sum_{\substack{f_j=1 \\ |f_i - f_j| = k}}^{m} P(f_i, f_j) \right\}$$

Energy : $$E = \sum_{f_i=1}^{m} \sum_{f_j=1}^{m} P(f_i, f_j)^2$$

where

$P(f_i, f_j)$ = number of pairs of connected components such that they are adjacent each other and their areas are f_i and f_j, respectively,

m = the maximum value of the aroa of the connected component.

Texture patterns in Fig. 11 were selected from the collection of natural textural materials and scenes in Ref. [8].

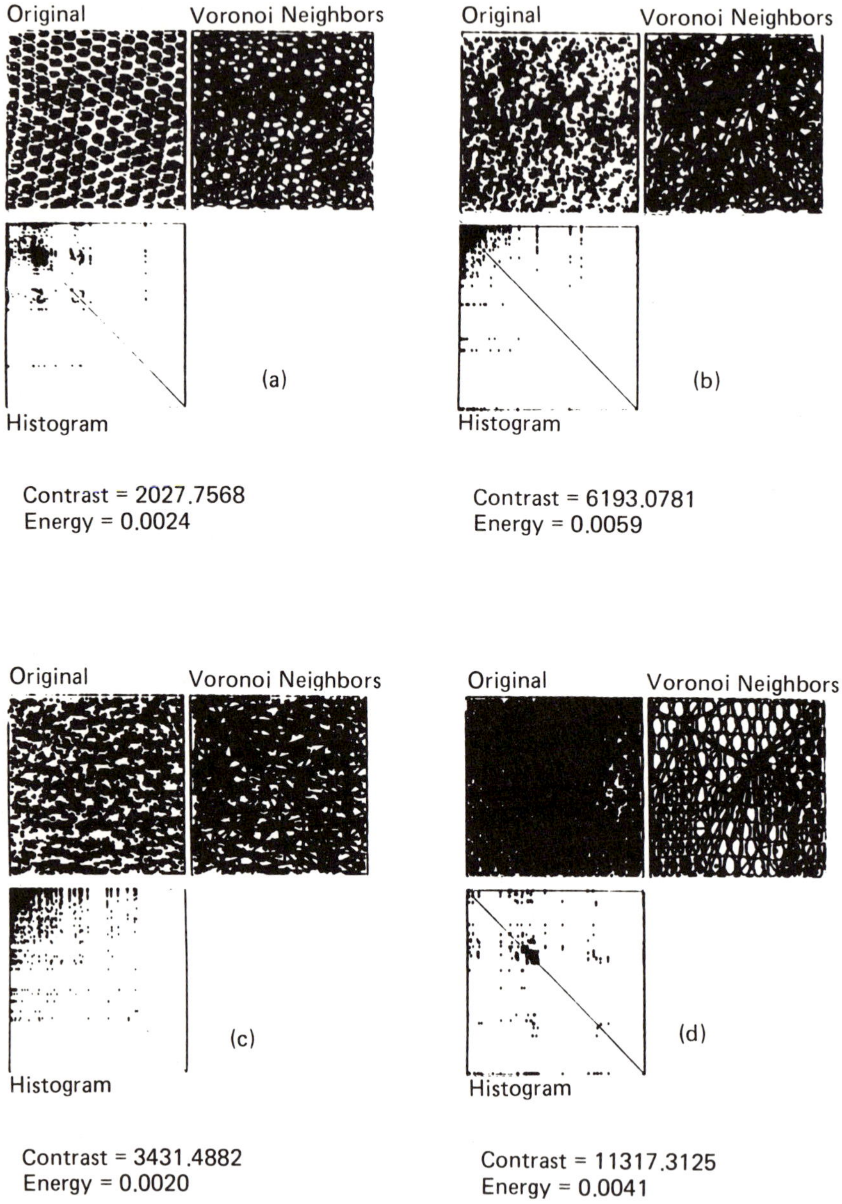

Fig. 11 Texture discrimination by cooccurrence matrix (top left: input, top right: adjacency graph, bottom left: cooccurrence matrix)

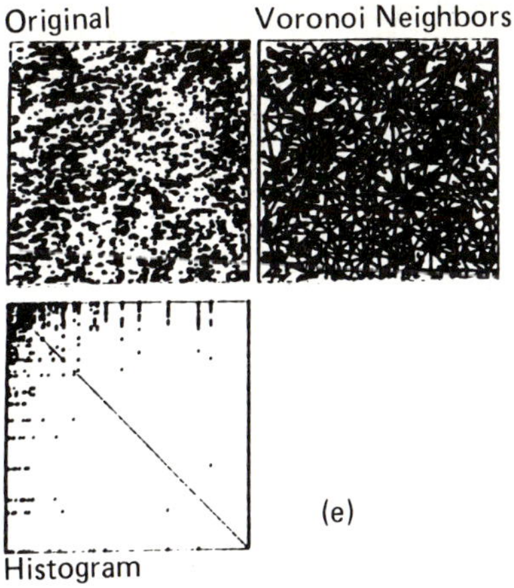

(e)

Contrast = 6158.6757
Energy = 0.0086

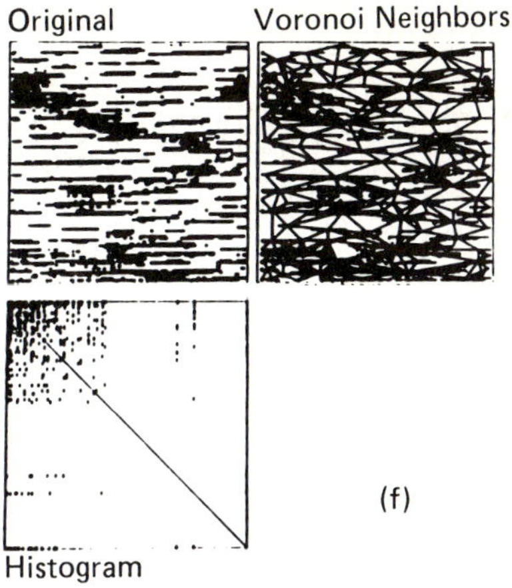

(f)

Contrast = 8785.8125
Energy = 0.0038

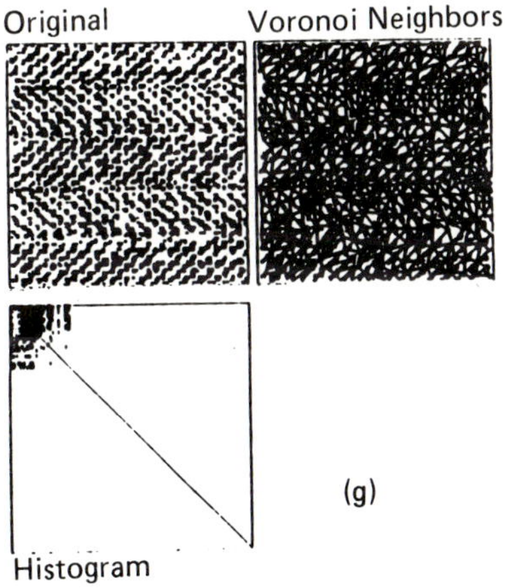

(g)

Contrast = 312.3818
Energy = 0.0020

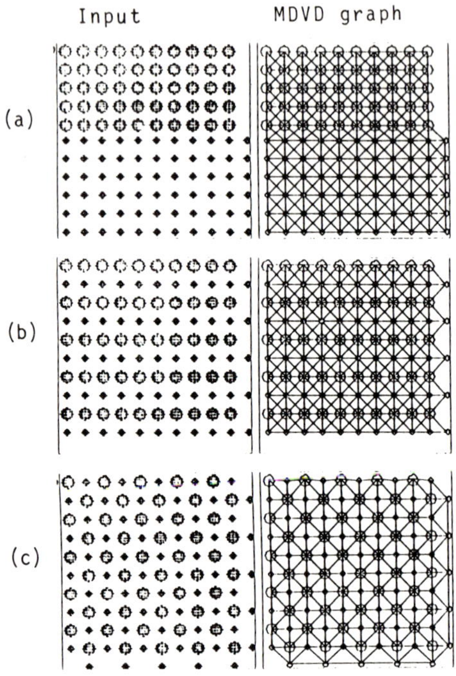

Fig. 12 Texture discrimination
left : original pattern, right : adjacency graph by MDVD,
feature : area of figure contrast : (a)94, (b)958, (c)832

7. Conclusion

In this paper we presented applications of the concepts of Dirichlet tessellation (DT) and the Voronoi diagram (VD), the relative neighbor, and the Gabriel neighbor in computational geometry to digital picture processing.

First, the above neighboring relations defined for a set of points in the continuous space were transferred to the digitized two-dimensional picture plane, and extended from a set of isolated pixels to a set of connected components, taking into consideration characteristics of digitized pictures.

Second, procedures to obtain the above neighboring relations for a given binary picture were discribed with estimation of the amount of computation required. Two kind of algorithms (parallel type and sequential type) were derived for obtaining the above modified digital Dirichlet tessellation (MDDT) and the modified digital Voronoi diagram (MDVD).

The adjacency graph induced from the neighboring relation is available in image processing, especially in texture analysis.

Feature vectors of individual connected components such as areas, lengths and other shape factors are given to the corresponding nodes, and those of component pairs such as distances between two neighboring components are asocciated with the graph edges.

Finally several examples of applications were presented. It is expected that the MDDT, the MDVD and the MDRN can be utilized in variety of fields of digital image processing such as triangulation of a figure, texture analysis of distributed connected components, description of particle-like figures (conversion into a graph description [9]), and data-dependent image segmentation.

Acknowledgements : Authors wish to thank Prof. Teruo Fukumura of Nagoya University for his encouragement. Parts of the work presented here were done by Mr. Kenji Mase and Mr. Yoshiyuki Yashima. Authors also thank Mr. Toyofumi Saito for preparing figures and typing the manuscript.

References

[1] G.T. Toussaint: "Pattern recognition and geometrical complexity", Proc. the 5th ICPR, pp.1324-1347 (1980)
[2] D.G. Kirkpatrick and J.D. Radke: "A framework for computational morphology", in G.T. Toussaint (ed.): Computational Geometry, Elsevier Science Pub. (North Holland) (1985)
[3] G.T. Toussaint: "The relative neighborhood graph of finite planar set", Pattern Recognition, 12, pp.261-268 (1980)
[4] J. Toriwaki, K. Mase, Y. Yashima and T. Fukumura: "Modified Voronoi diagram and relative neighbors on a digitized picture and their applications to tissue image analysis", Proc. the 1st ISMIII, pp.262-267 (Oct. 1982)
[5] K. Mase, J. Toriwaki and T. Fukumura: "Modified digital Voronoi diagram and its applications to image processing", Trans. IECE, Japan, J64-D, pp.1029-1036 (Nov. 1981)
[6] J. Toriwaki and S. Yokoi: "Distance transformations and skeletons of digitized pictures with applications", in A. Rosenfeld and L. Kanal (eds.): Progress in Pattern Recognition, 1, pp.187-264, North Holland, N.Y., 1981
[7] R.M. Haralick, K.S. Shanmugan and I. Dinstein: "Texture features for image classification", IEEE Tr. SMC, SMC-3, 6, pp.610-621 (Nov. 1973)
[8] P. Brodatz: Texture, Dover (1966)
[9] J.M.S. Prewitt: "Contemporary medical microscopy: the advent of intelligent microscope", IEEE Trans. NS, NS-27, 3, pp.1207-1217 (June 1980)
[10] C.T. Zahn and R.Z. Roskies: "Fourier descriptors for plane closed curves", IEEE Trans. on Computers, C-21, pp.269-281 (1972)

Appendix

We will present here an alternative form of the definition for the MDDT [5].

[Definition A.1] Given two pixels $\underline{x} = (r,s)$ and $\underline{y} = (p,q)$, half-planes $H\underline{xy}$ and $H\underline{yx}$ are defined as follows.

$$\underline{z} \in H\underline{xy}, \text{ if } d_t(\underline{x},\underline{z}) < d_t(\underline{y},\underline{z})$$

$$\underline{z} \in H\underline{yx}, \text{ if } d_t(\underline{x},\underline{z}) > d_t(\underline{y},\underline{z})$$

where the distance $d_t(\underline{x},\underline{y})$ between pixels $\underline{x}$ and $\underline{y}$ is defined by

either of
(1) and (2) below[+].
(1) $t = 8$, if $|r - p| = |s - q|$, and otherwise $t = 4$.
(2) $t = 4$, if $r = p$ or $s = q$, and otherwise $t = 8$.

This means that, for the example shown in Fig. 1, we conveniently use the perpendicular bisector between the pixels Ⓐ and Ⓑ as the border between two tiles corresponding to those pixels.

[Definition A.2] Given two connected components C1 = {$\underline{x}1$, $\underline{x}2$, ···, $\underline{x}n1$} and C2 = {$\underline{y}1$, $\underline{y}2$, ···, $\underline{y}n2$}, two half-planes H_{12} and H_{21} are defined as follows.

$$H_{12} = \bigcup_{i=1}^{n} \left(\bigcap_{j=1}^{n} H_{\underline{x}_i \underline{y}_j} \right), \qquad H_{21} = \bigcup_{j=1}^{n_2} \left(\bigcap_{i=1}^{n_1} H_{\underline{y}_j \underline{x}_i} \right)$$

where $H_{\underline{x}_i \underline{y}_j}$ and $H_{\underline{y}_j \underline{x}_i}$ are half-planes defined by Definition A.1 for a pixel pair $\underline{x}_i$ and $\underline{y}_j$.

[Definition A.3] For a set of n connected components C1, C2, ···, Cn,

$$Tk = \bigcap_{j \neq k} H_{kj}$$

is called the tile corresponding to the connected component Ck. The whole of T1, T2, ···, Tn is called the MDDT.

Although the MDDT and MDVD defined here do not coincide strictly with those by Definitions 3 and 4 in the paper, difference among them is negligible from the practical view point.

Computation time increases remarkably when the above definition is adopted. For an input picture of 30 x 40 pixels, for example, it took from 120 to 180 sec. to determine the MDVD according to Definitions A.2 and A.3 while only about 0.2 sec. by using Definition 4 in the text. In fact, if we use Definition A.2 and A.3, we must generate $H_{\underline{x}\underline{y}}$ and $H_{\underline{y}\underline{x}}$ for all pairs of $\underline{x}_i$, $\underline{y}_j$ for each pair of connected components C_i and C_j. This process must be repeated for all of k and j's to obtain Tk.

[+] Although axioms of the metric are not satisfied for this distance function, a half-plane can be determined for an arbitrary pair of two pixels using the perpendicular bisector between those two pixels. Thus no problem is caused in the following definitions.

Computational Morphology
G.T. Toussaint (editor)

A GRAPH-THEORETICAL PRIMAL SKETCH

Godfried T. Toussaint

School of Computer Science
McGill University
Montreal, Canada

ABSTRACT
A new graph termed the *sphere-of-influence* graph is proposed as a *primal sketch* intended to capture the low-level perceptual structure of visual scenes consisting of dot patterns. This graph suffers from none of the serious drawbacks of previous methods and for a pattern consisting of n dots, can be computed efficiently in O(n log n) time.

1. INTRODUCTION

The central problem in computer vision begins with a grey-level intensity array of a visual pattern or scene and culminates with a description which is a marriage of the array itself (the bottom-up or data-driven information) and the purpose for which it is viewed (the top-down or conceptually-driven information). Considerable attention has been devoted to low-level vision, i.e., the aspect dealing with the analysis of the intensity array up to the level of figure-ground separation. The main purpose of analyzing an intensity array at this level is to form a low-level description that is independent of, or holds true for, any final conceptually-driven description. David Marr [1] has called such a description a *primal sketch* of the intensity array. One class of patterns or scenes which have been studied extensively is the so-called *dot-pattern.*

When points in the plane have a finite diameter so that they are visible, and when they are fairly densely and uniformly distributed in some region in the plane then a human observer is quick to perceive the "shape" of such a set. These sets are usually referred to as *dot patterns* or *dot figures.* A polygonal description of the boundary of the shape is referred to as the *shape hull* of a dot pattern, where the vertices are given in terms of the cartesian coordinates of the centers of the dots. There are two versions of the shape hull problem: in one there are no "holes" in the dot pattern and the dot pattern is "simply connected" and hence the shape hull is a simple polygon, whereas in the more difficult problem both "holes" and "disconnected" components may exist. To add to this difficulty, in some instances illusory contours are perceived between "disconnected" components as illustrated by Kennedy and Ware [2]. For more details on this problem and early approaches to solving it, see [3]. A more recent paper addressing this problem is that of Medek [4].

In addition to describing the shape or structure of a set of points by its *shape-hull* or *external shape* we may also use the *skeleton* or *internal shape.* An early step in this direction was taken by Zahn [5] with the minimal spanning tree. More recent approaches have used the *relative neighborhood graph* [6] - [12] and [24] - [26].

Most of the early methods suffer from various deficiencies, most notably computational inefficiency, lack of elegance, or dependence on various parameters which must be fine-tuned in order to obtain satisfactory performance on the task at hand.

A very elegant definition of the external shape of a set of points was put forward recently by Edelsbrunner, Kirkpatrick, and Seidel [13]. They propose a natural generalization of convex hulls that they call α-hulls. The α-hull of a point set is based on the notion of generalized discs in the plane. For arbitrary real valued α, a *generalized disc of radius* $1/\alpha$ is defined as follows [13]:

(i) if $\alpha>0$, it is a (standard) disc of radius $1/\alpha$;

(ii) if $\alpha<0$, it is the complement of a disc of radius $1/\alpha$; and

(iii) if $\alpha=0$, it is a half plane.

The α-*hull* of a point set S is defined to be the intersection of all closed generalized discs of radius $1/\alpha$ that contain all the points S. The convex hull of S is precisely the 0-hull of S. The family of α - hulls includes the smallest enclosing circle (when $\alpha=1/\text{radius (S)}$), the set S itself (for all α sufficiently small) and an essentially continuous set of enclosing regions in between these extremes.

Edelsbrunner *et al* [13] also define a combinatorial variant of the α-hull, called the α-*shape* of a point set, which can be viewed without serious misunderstanding as the boundary of the α-hull with curved edges replaced by straight edges. Unlike the family of α-hulls, the family of distinct α-shapes has only finitely many members. These provide a spectrum of progressively more detailed descriptions of the external shape of a given point set. In [13], efficient algorithms are also presented for computing the α-shapes of dot patterns consisting of n points in 0(n log n) time. Subsequently, Kirkpatrick and Radke [16] outlined a new methodolgy for describing the *internal shape* of planar point sets. We should note that the ideas in [13] are closely related to the notions of *opening* and *closing* sets, found in mathematical morphology [14], [15]. See also [17].

While the α-shape is elegant and efficient to compute, it still suffers from two drawbacks: a single parameter, namely α, must nevertheless be tuned, and the final α-shape is limited to describing the shape of a "single" pattern rather than a scene of several disconnected objects.

One of the very few conceptually simple and elegant methods, for extracting the shape of a dot pattern, that is free of parameter tuning is due to Rosenberg and Langridge [18]. Their algorithm, however, suffers from at least two drawbacks. First, it renders only a single object and, secondly, it is computationally inefficient. Although no complexity analysis is given in [18], it is easy to show that their algorithm runs in $0(n^4)$ time.

II. THE SPHERE-OF-INFLUENCE GRAPH

In this section we propose a graph that appears to capture the essence of a *primal sketch* for dot patterns of arbitrary complexity. The graph suffers from none of the drawbacks of previous methods: 1) it delivers either the "internal" structure in the form of a "skeleton" or the "external" structure in the form of a "shape-hull" as a function of what the data look like, 2) it outputs either a connected graph or a collection of disconnected pieces where appropriate, 3) it performs all of the above "completely automatically", i.e., without any requisite tuning of parameters, and 4) it can be computed efficiently.

We call this graph the *sphere-of-influence* graph. In addition to the above properties, the graph affords a graph-theoretical explanation of some visual illusions such as the Mueller-Lyer illusion.

Let $S=\{x_1,x_2, \ldots, x_n\}$ be a finite set of points in a plane. For each point $x \in S$, let r_x be the closest distance to any other point in the set, and let C_x be the circle of radius r_x centered at x. The sphere of influence graph is a graph on S with an edge between points x and y if and only if the circles C_x and C_y intersect in at least two places. It is shown in [19] that:

(i) The sphere of influence graph has at most 29n edges ($n=|S|$).

(ii) Every decision tree algorithm for computing the sphere of influence graph requires at least Ω(n log n) steps in the worst case.

As an application of (i), El Gindy [20] observed that an algorithm of Bentley and Ottman [21] can be used to find the sphere of influence graph of S (denoted by SIG (S)) in 0(n log n) time.

Figures 1-50 illustrate sets of points (upper half of the page) along with their sphere-of-influence graphs (lower half of the page).

The purpose of presenting this variety of instantiations of the SIG (S) is to give the reader a chance to browse and ponder for him-or-her self on the power of this graph as a primal sketch. Hence we leave out an in-depth discussion of possible weaknesses in certain dot figures. Hopefully, the readers will know more than the writer about their own pet applications problems where the SIG may be helpful. It should be remembered that the SIG is a low-level vision operator and thus could be "plugged-in" to any higher-level vision program the reader may care to design.

We do mention in passing, however, that as Figures 31-35 illustrate, the SIG (S) does provide a graph-theoretical "explanation" of the dot-pattern version of the Mueller-Lyer illusion. A detailed treatment of this illusion in both its line-segment and dot-pattern forms can be found in [22]. Referring to Figures 31 and 32, if we measure the length of the central vertical line connecting the portions of the SIG(S) where they branch out, we see that the line segment so obtained is indeed shorter in Fig. 32 than in Fig. 31. Furthermore, if the angle of the "arrows" is made less acute as in Figs. 33 and 34, the magnitude of the "illusion" decreases in agreement with empirical evidence [22].

We close by mentioning two more structures closely related to the SIG which may find useful application in shape analysis.

Motivated by the work in [13] we can also use the *sphere-of-influence* graph to define the *boundary* of a planar set S as either the *contour* of the union of the circles C_x, $x \in S$ (the *sphere-of-influence hull*) or the graph composed of those edges corresponding to pairs of points in S that have adjacent arcs on the *contour* defined above (the *sphere-of-influence shape*). Figures 51-54 illustrate two sets of points and their sphere-of-influence hull. These structures may be interpreted as smoothed or blurred versions of the original, that depend nonuniformly on the local structure of the data. While the sphere-of-influence hull can be computed from the SIG, it is interesting to note that it can also be computed in O(n log n) time without first computing the SIG, by using the Laguerre Voronoi diagram [23]. The algorithm in [23] computes the boundary or contour of the union of n arbitrary circles in O(n) time from their Laguerre Voronoi diagram. Thus to compute the sphere-of-influence hull all we need to do is to construct the circles C_{x_i}, i=1,2,...,n, for which we must compute for each $x_i \in s$, its nearest neighbour in s. This can be done in O(n log n) time with the ordinary Voronoi diagram [3].

ACKNOWLEDGEMENT

The author is grateful to Teren Gum for programming the SIG algorithm and plotting the results. Research is supported by NSERC grant No. A9293, FCAR grant No. EQ - 1678, and a Killam Senior Research Fellowship awarded by the Canada Council.

REFERENCES

[1] D. Marr, "Early processing of visual information", *Philosophical Transactions of the Royal Society of London*, B. Biological Sciences, Vol. 275, No. 942, pp. 483-524, October 1976.

[2] J. M. Kennedy & C. Ware, "Illusory contours can arise in dot figures", *Perception*, Vol. 7, 1978, pp. 191-194.

[3] G. T. Toussaint, "Pattern recognition and geometrical complexity", *Proc. 5th International Conference on Pattern Recognition*, Miami Beach, December 1980, pp. 1324-1347.

[4] V. Medek, "On the boundary of a finite set of points in the plane", *Computer Graphics & Image Processing*, vol. 15, 1981, pp. 93-99.

[5] C. T. Zahn, "Graph-theoretical methods for detecting and describing gestalt clusters", *IEEE Trans. Computers*, Vol. C-20, Jan. 1971, pp. 68-86.

[6] G. T. Toussaint, "The relative-neighbourhood graph of a finite planar set", *Pattern Recognition*, Vol. 12, 1980, pp. 261-268.

[7] J. O'Rourke, "Computing the relative-neighbourhood graph in the L_1 and L_∞ metrics", *Pattern Recognition*, Vol. 15, 1982, pp. 189-192.

[8] R. B. Urquhart, "Some properties of the planar Euclidean relative neighbourhood graph", *Pattern Recognition Letters*, Vol. 1, 1983, pp. 317-322.

[9] K. J. Supowit, "The relative neighbourhood graph with an application to minimum spanning trees", *J.A.C.M.*, Vol. 30, 1983, pp. 428-448.

[10] D. T. Lee, "Relative neighbourhood graphs in the L_1 metric", *Pattern Recognition*, Vol. 18, 1985, pp. 327-332.

[11] M. Ichino & J. Sklansky, "The relative neighbourhood graph for mixed feature variables", *Pattern Recognition*, Vol. 18, 1985, pp. 161-167.

[12] J. Katajainen & O. Nevalainen, "Computing relative neighbourhood graphs in the plane", *Pattern Recognition*, Vol. 19, No. 3, 1986, pp. 221-228.

[13] E. Edelsbrunner, D. G.Kirkpatrick & R. Seidel, "On the shape of a set of points in the plane", *IEEE Trans. Information Theory*, Vol. 29, July 1983, pp. 551-559.

[14] G. Matheron, *Random Sets and Integral Geometry*, John Wiley, 1975.

[15] J. Serra, *Image Analysis & Mathematical Morphology*, Academic Press, 1982.

[16] D. G. Kirkpatrick & J. D. Radke, "A framework for computational morphology" in *Computational Geometry*, G. T. Toussaint, Ed., North Holland, 1985, pp. 217-248.

[17] G. T. Toussaint, "Computational geometry & morphology", *Proc. First International Symposium for Science on Form*, Tsukuba, Japan, November 1985, pp. 395-403.

[18] B. Rosenberg & D. J. Langridge, "A computational view of perception", *Perception*, Vol. 2, 1973, pp. 415-424.

[19] D. Avis & J. Horton, "Remarks on the sphere of influence graph", in *Discrete Geometry and Convexity*, Eds., J. E. Goodman et al., New York Academy of Sciences, 1985, pp. 323-327.

[20] Hossam El Gindy, personal communication.

[21] J. Bentley & T. Ottman, "Algorithms for reporting and counting geometric intersections", *IEEE Trans. Computers*, Vol. C-28, pp. 643-647.

[22] S. Coren & J. S. Girgus, *Seeing is Deceiving: The Psychology of Visual Illusions*, Lawrence Erelbaum Associates, 1978.

[23] H. Imai, M. Iri, & K. Murota, "Voronoi diagram in the Laguerre geometry and its applications", *SIAM Journal of Computing*, Vol. 14, No. 1, 1985, pp. 93-105.

[24] G. Toussaint & R. Menard, "Fast algorithms for computing the planar relative neighbourhood graph", in *Methods of Operations Research*, University of Koln, August 1980, pp. 425-428.

[25] R. B. Urquhart, "Algorithms for computation of relative neighbourhood graph", *Electronics Lettes*, Vol. 16, 1980, No. 14, pp. 556-557.

[26] G. T. Toussaint, "Algorithms for computing relative neighbourhood graph", *Electronics Letters*, Vol. 16, 1980, No. 22, pp. 860-861.

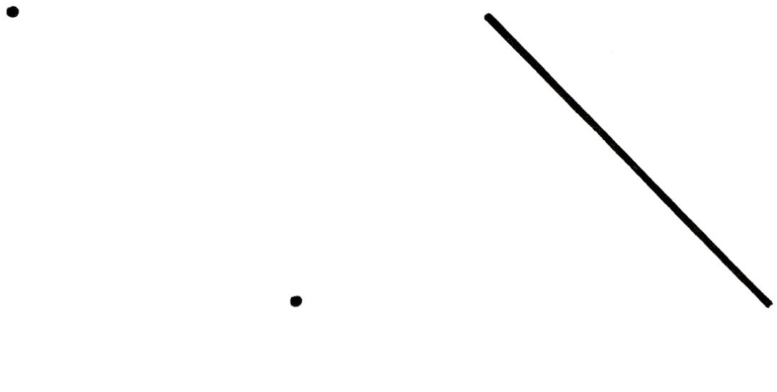

FIGURE 1. NUMBER OF POINTS : 2.

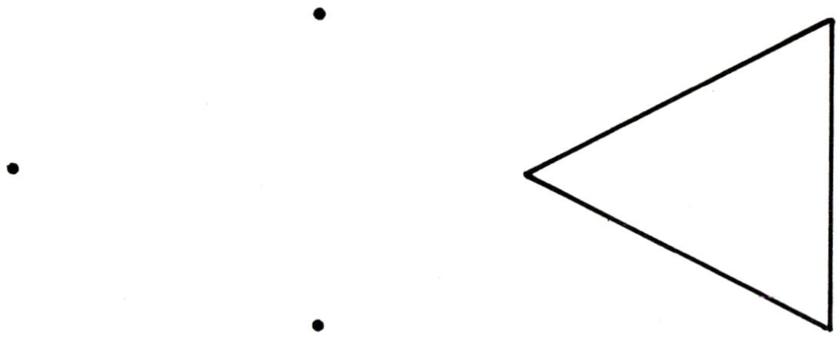

FIGURE 2. NUMBER OF POINTS : 3.

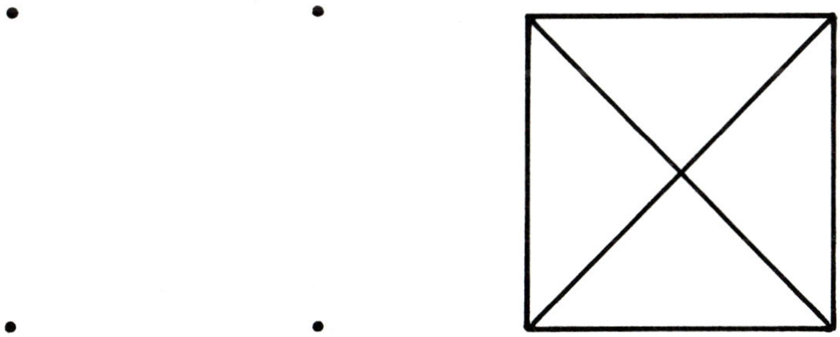

FIGURE 3. NUMBER OF POINTS : 4.

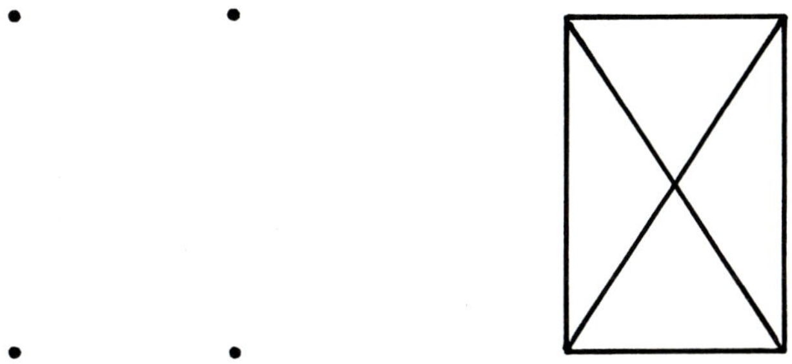

FIGURE 4. NUMBER OF POINTS : 4.

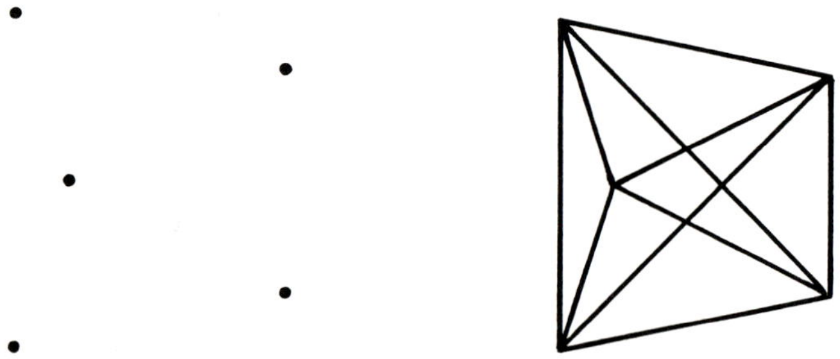

FIGURE 5. NUMBER OF POINTS : 5.

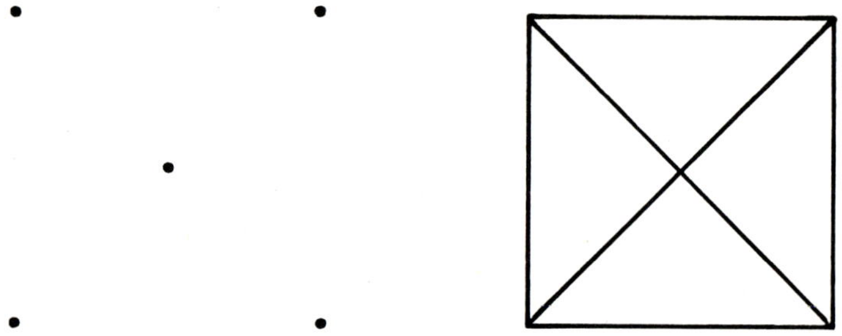

FIGURE 6. NUMBER OF POINTS : 5.

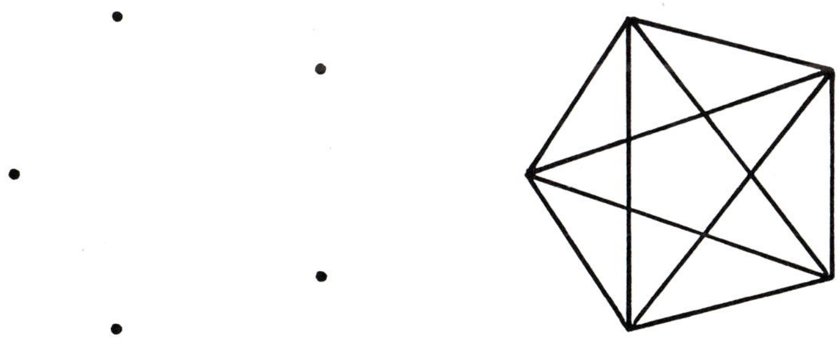

FIGURE 7. NUMBER OF POINTS : 5.

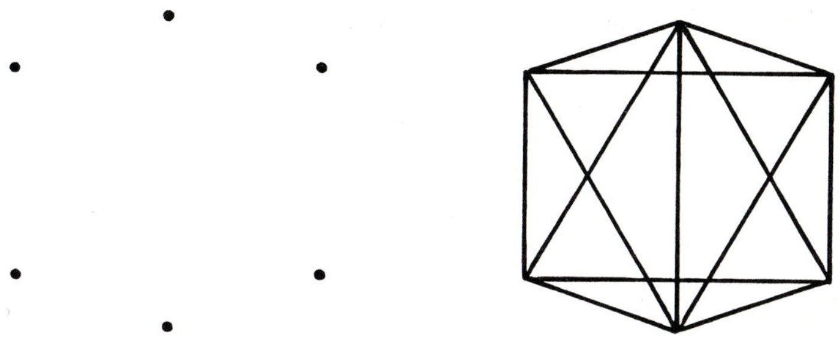

FIGURE 8. NUMBER OF POINTS : 6.

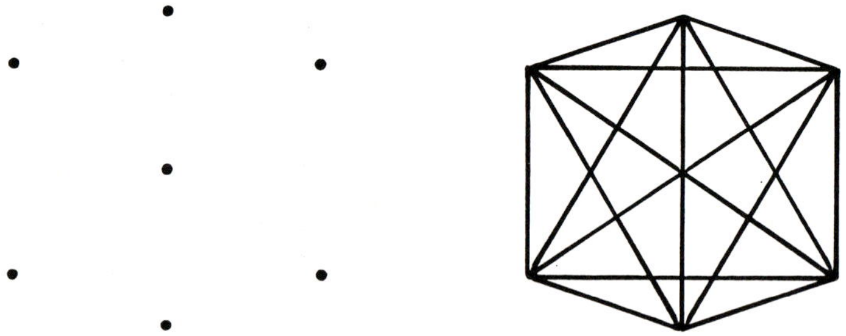

FIGURE 9. NUMBER OF POINTS : 7.

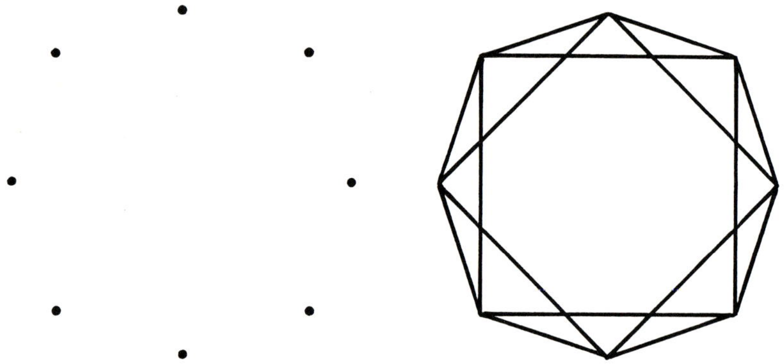

FIGURE 10. NUMBER OF POINTS : 8.

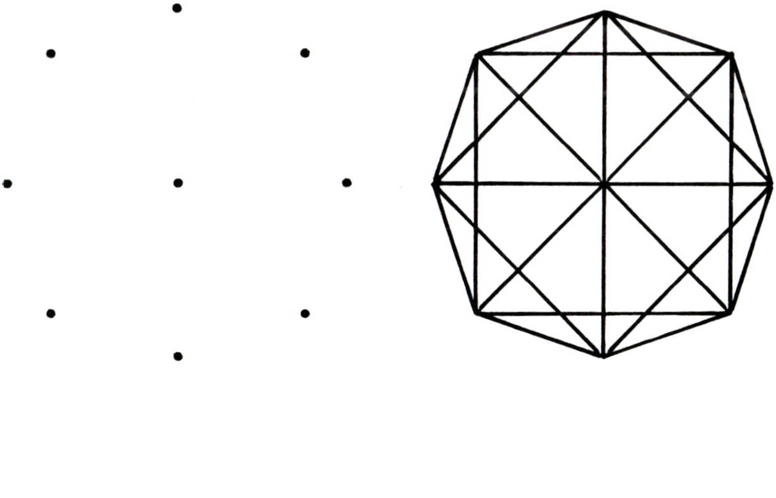

FIGURE 11. NUMBER OF POINTS : 9.

FIGURE 12. NUMBER OF POINTS : 8.

FIGURE 13. NUMBER OF POINTS : 8.

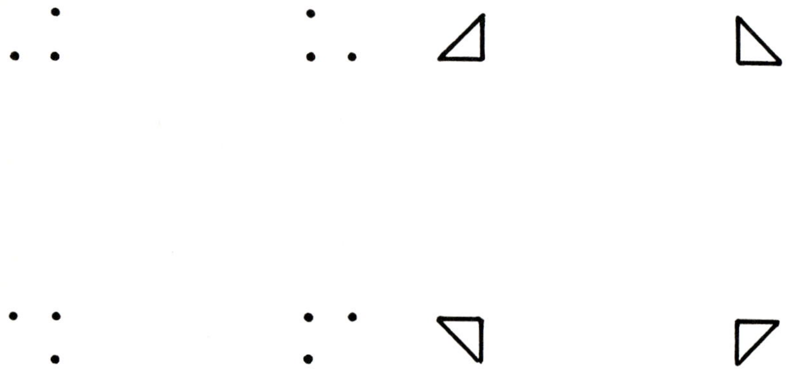

FIGURE 14. NUMBER OF POINTS : 12.

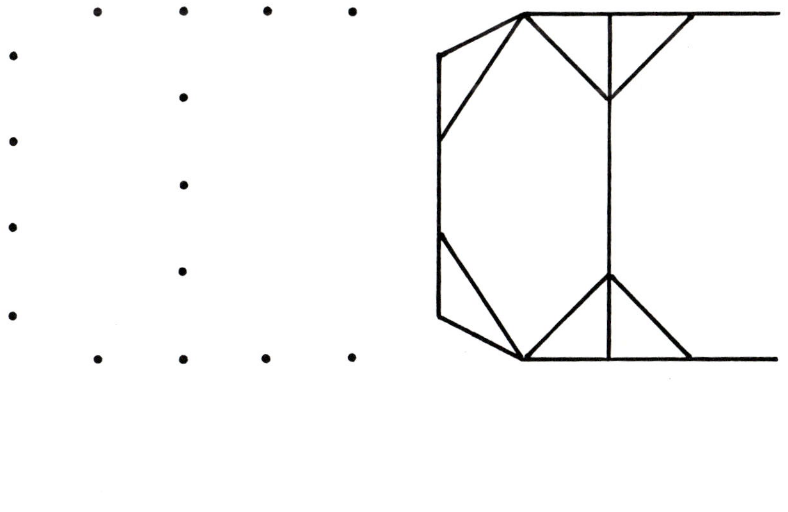

FIGURE 15. NUMBER OF POINTS : 15.

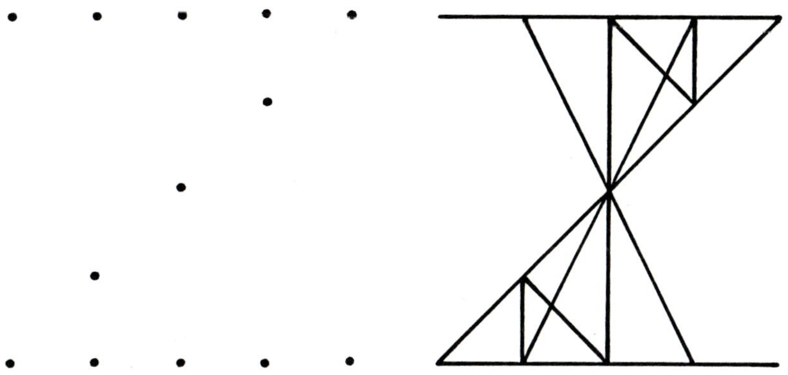

FIGURE 16. NUMBER OF POINTS : 13.

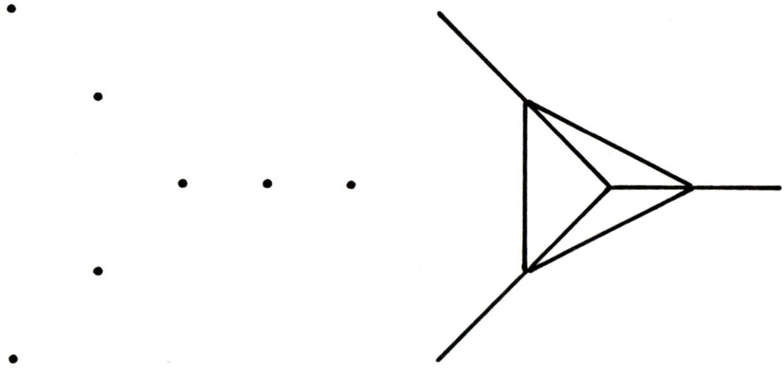

FIGURE 17. NUMBER OF POINTS : 7.

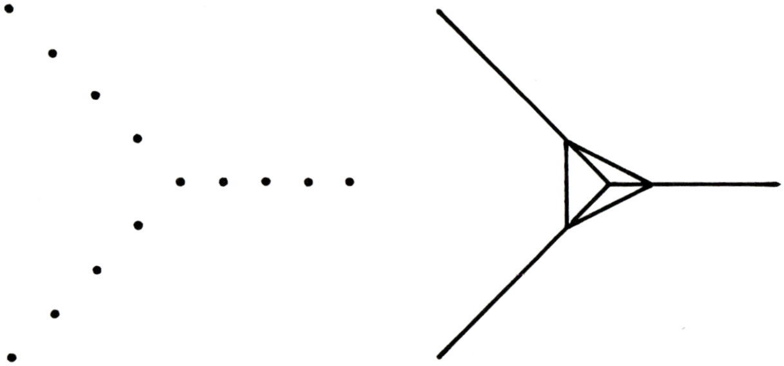

FIGURE 18. NUMBER OF POINTS : 13.

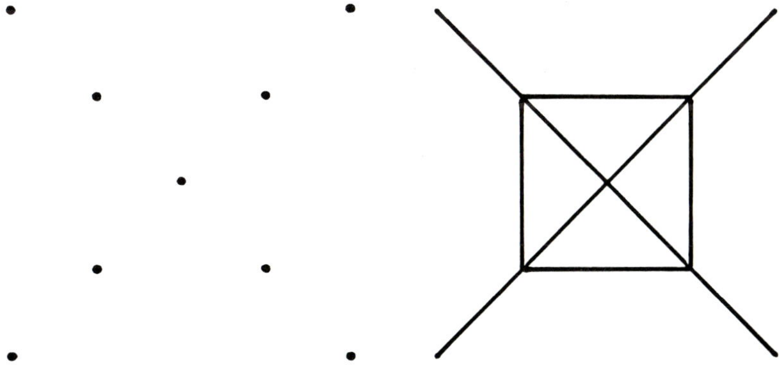

FIGURE 19. NUMBER OF POINTS : 9.

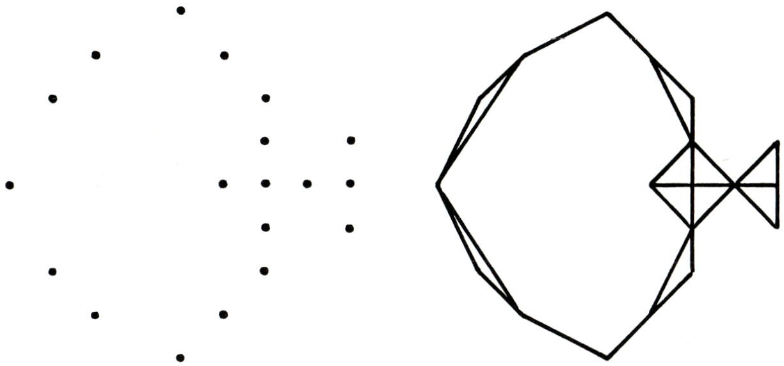

FIGURE 20. NUMBER OF POINTS : 19.

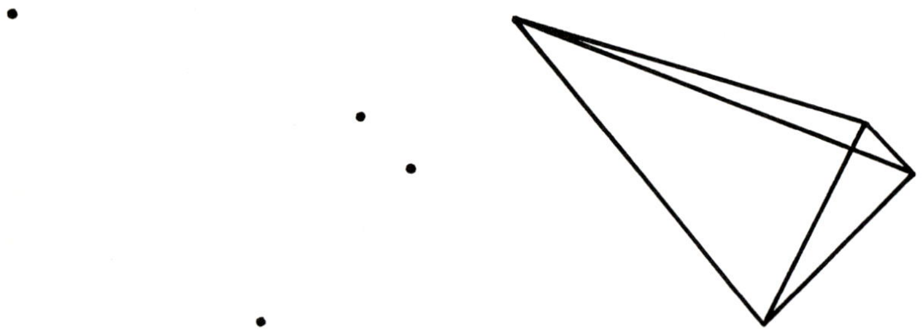

FIGURE 21. NUMBER OF POINTS : 4.

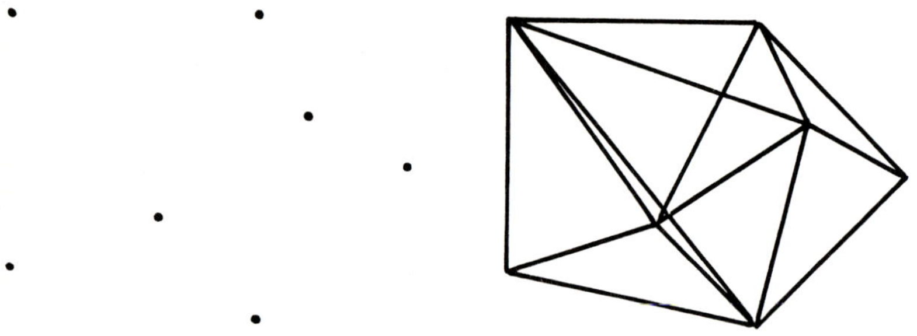

FIGURE 22. NUMBER OF POINTS : 7.

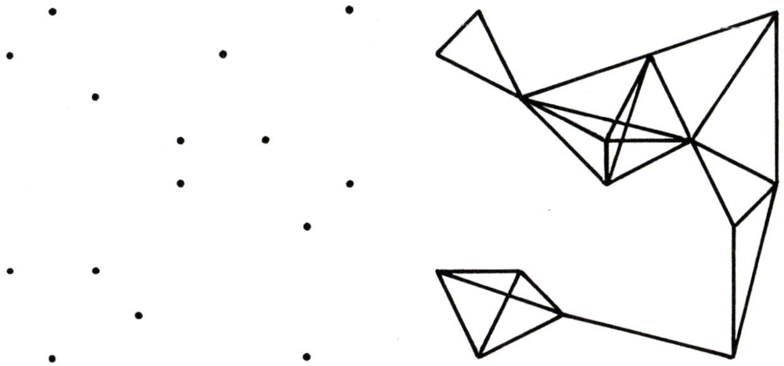

FIGURE 24. NUMBER OF POINTS : 15.

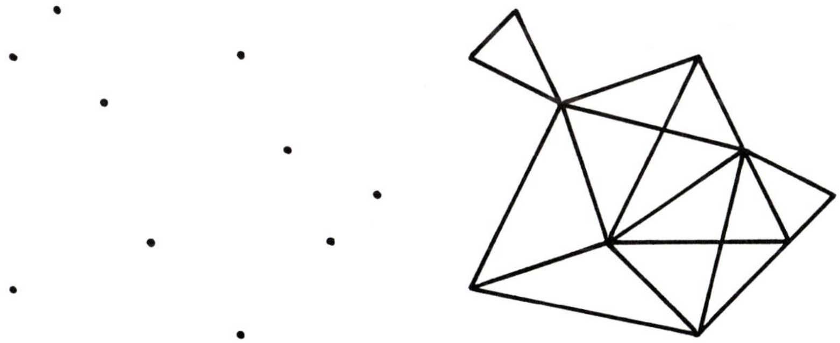

FIGURE 23. NUMBER OF POINTS : 10.

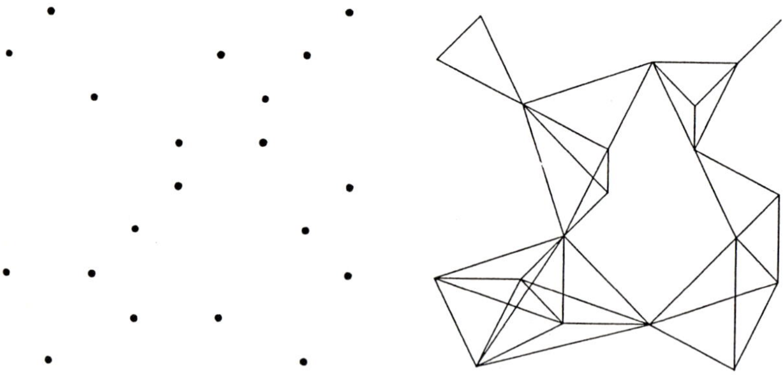

FIGURE 25. NUMBER OF POINTS : 20.

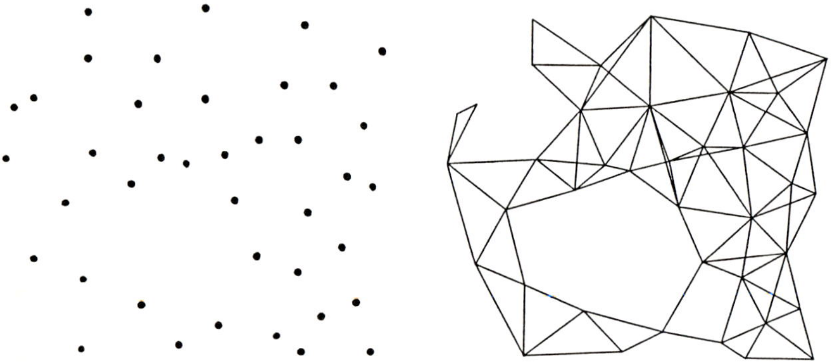

FIGURE 26. NUMBER OF POINTS : 40.

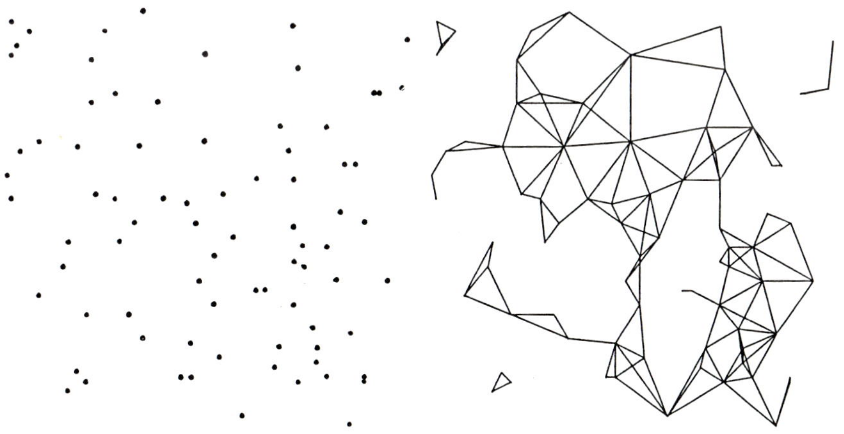

FIGURE 27. NUMBER OF POINTS : 80.

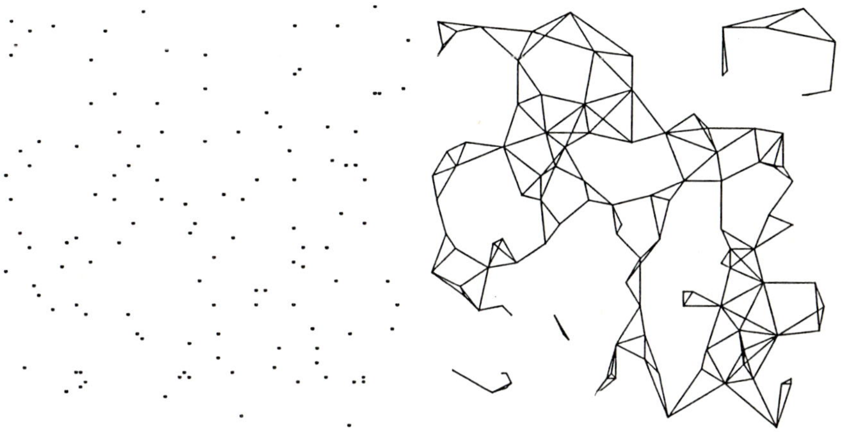

FIGURE 28. NUMBER OF POINTS : 120.

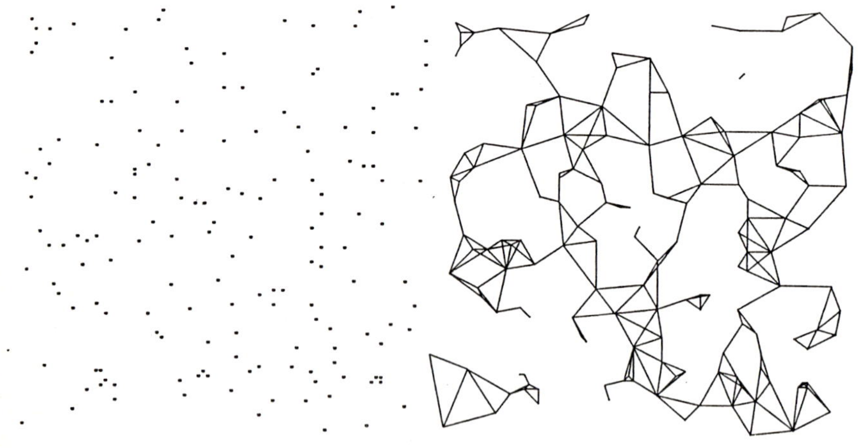

FIGURE 29. NUMBER OF POINTS : 160.

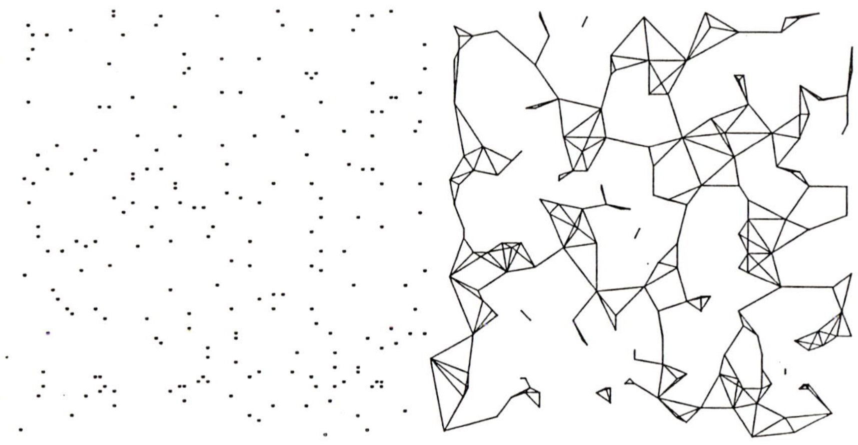

FIGURE 30. NUMBER OF POINTS : 200.

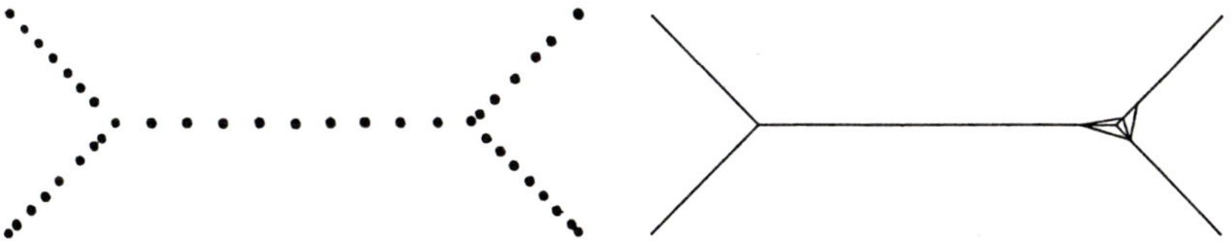

FIGURE 31. NUMBER OF POINTS : 40.

FIGURE 32. NUMBER OF POINTS : 38.

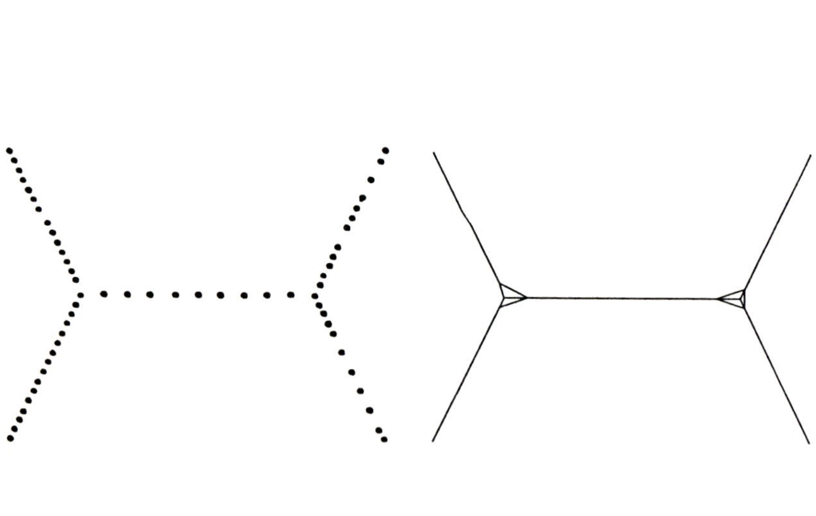

FIGURE 33. NUMBER OF POINTS : 62.

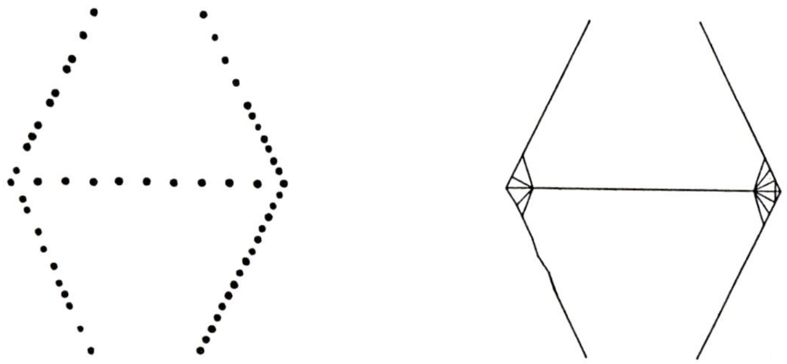

FIGURE 34. NUMBER OF POINTS : 57.

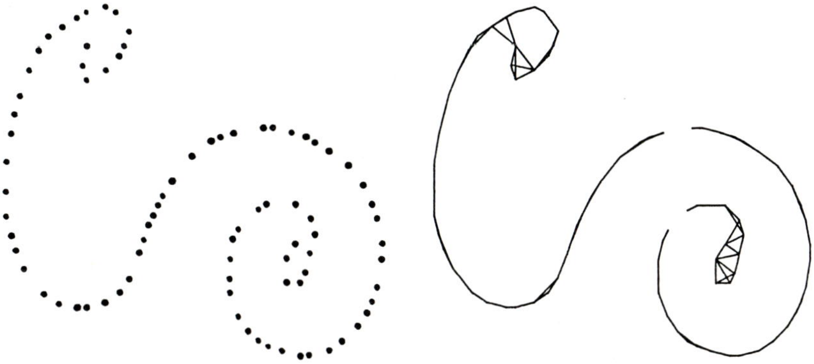

FIGURE 36. NUMBER OF POINTS : 83.

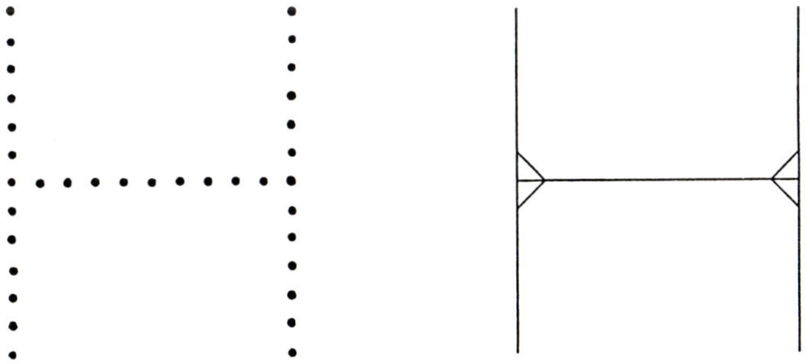

FIGURE 35. NUMBER OF POINTS : 35.

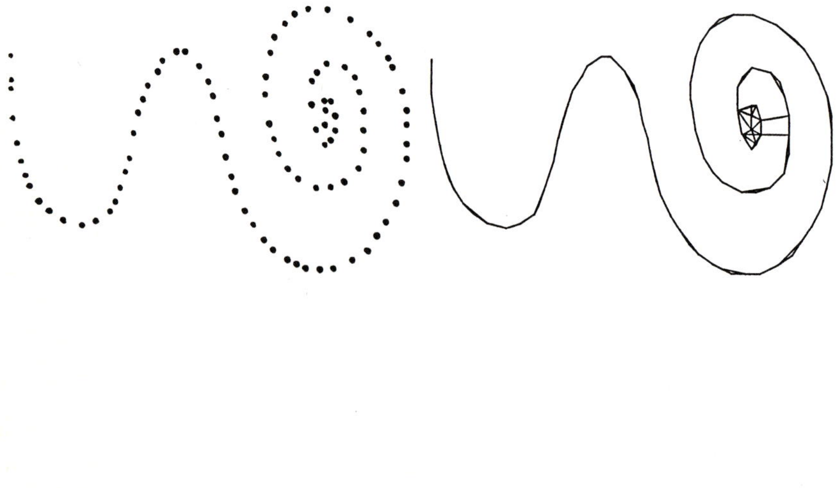

FIGURE 37. NUMBER OF POINTS : 95.

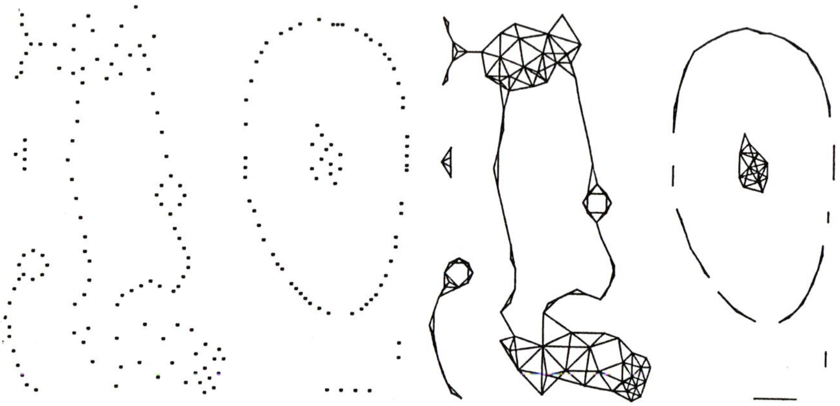

Figure 38

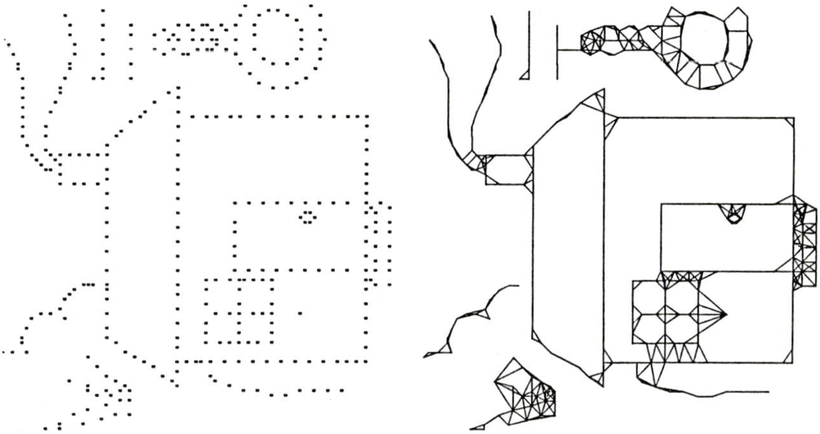

FIGURE 40. NUMBER OF POINTS : 354.

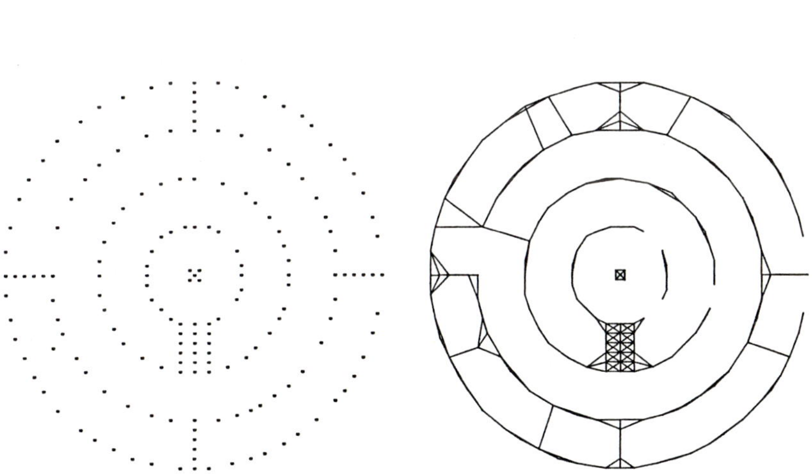

FIGURE 39. NUMBER OF POINTS : 181.

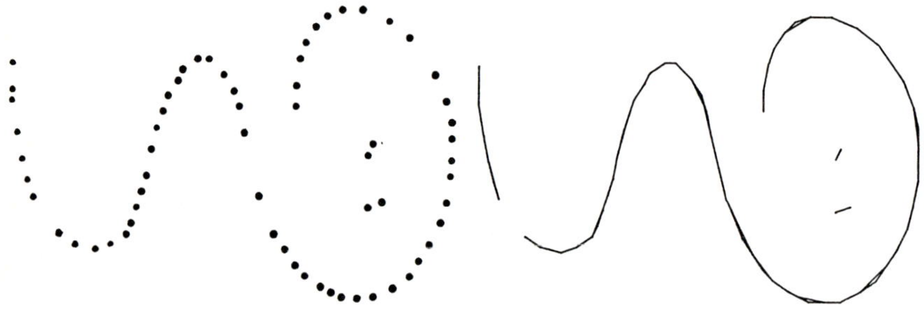

FIGURE 42. NUMBER OF POINTS : 64.

FIGURE 41. NUMBER OF POINTS : 56.

FIGURE 43. NUMBER OF POINTS : 132.

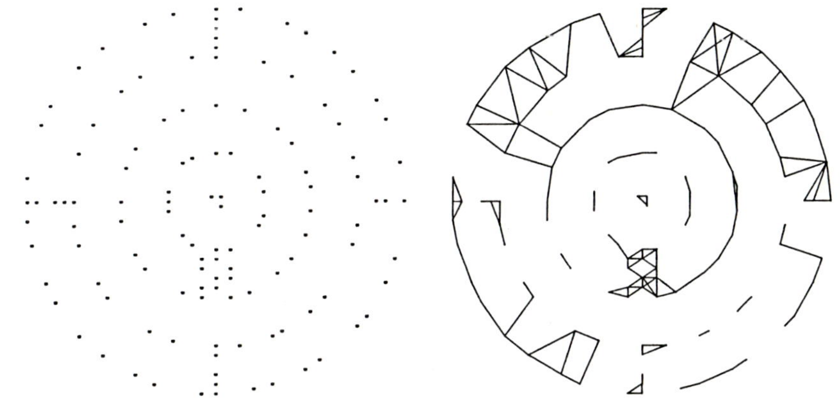

FIGURE 44. NUMBER OF POINTS : 123.

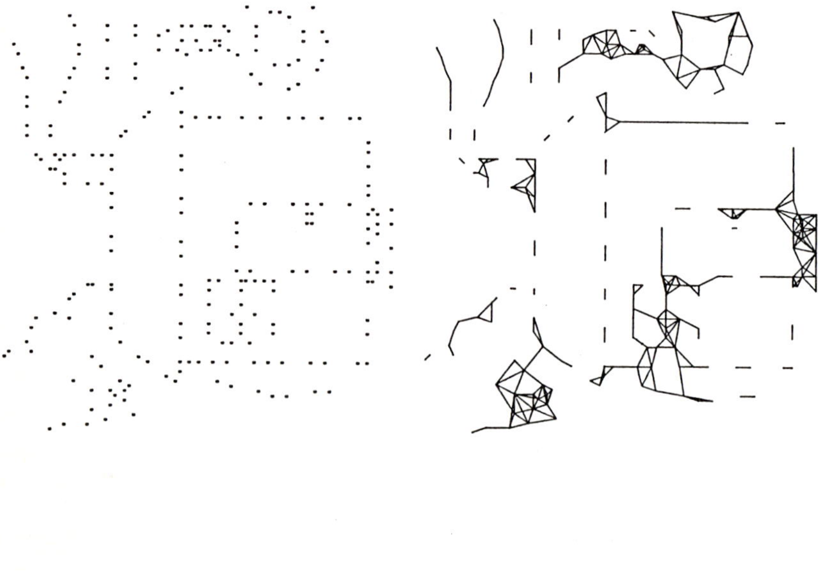

FIGURE 45. NUMBER OF POINTS : 239.

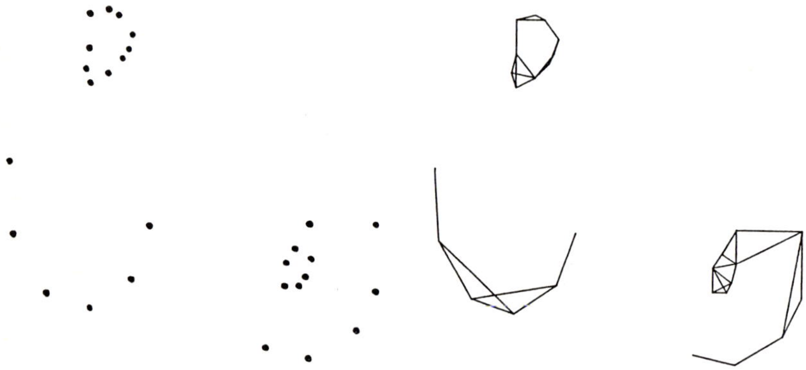

FIGURE 46. NUMBER OF POINTS : 28.

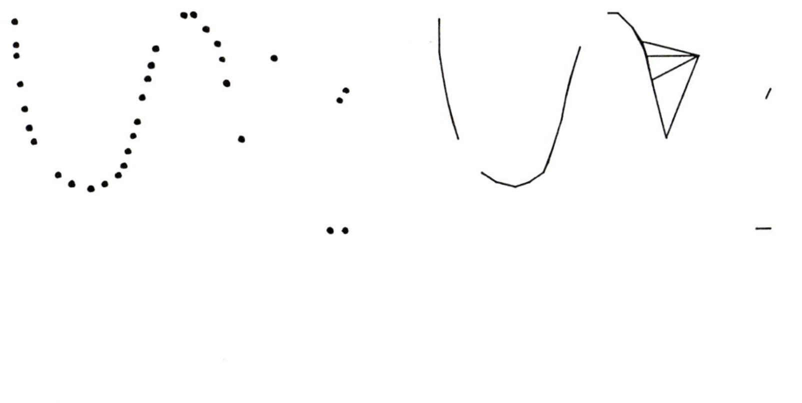

FIGURE 47. NUMBER OF POINTS : 32.

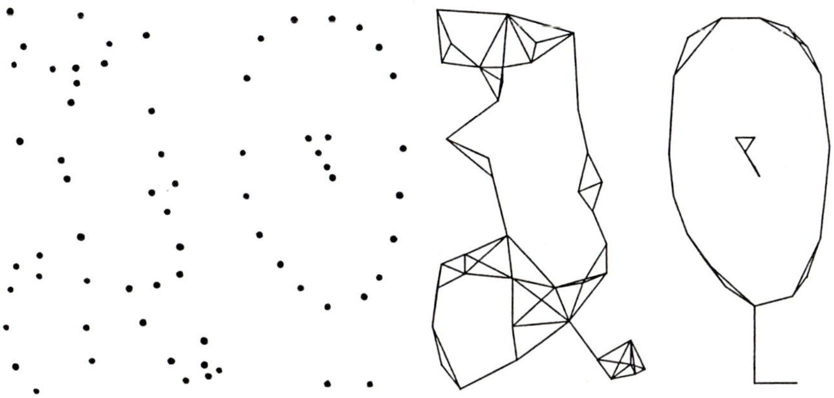

FIGURE 48. NUMBER OF POINTS : 66.

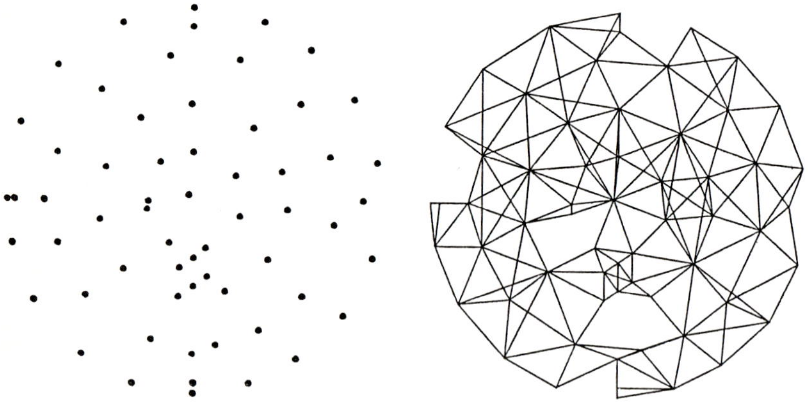

FIGURE 49. NUMBER OF POINTS : 61.

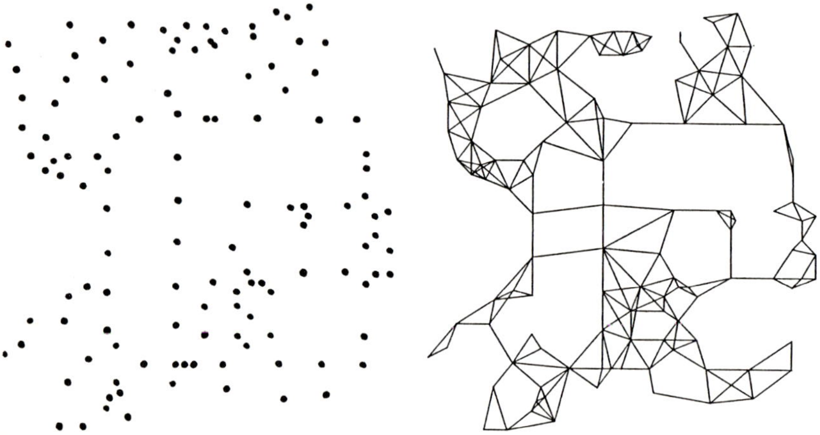

FIGURE 50. NUMBER OF POINTS : 121.

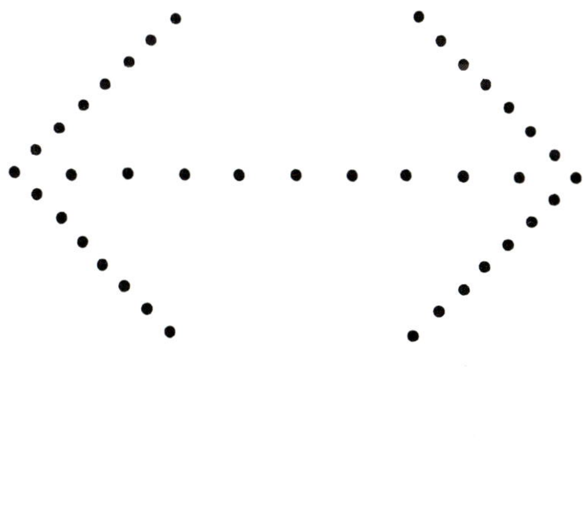

Figure 51: A set of points.

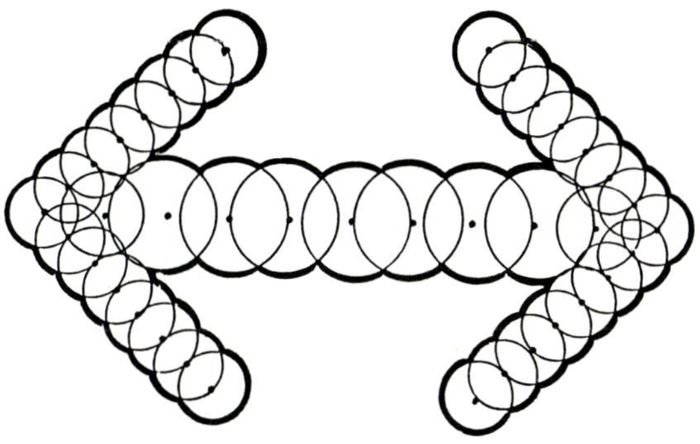

Figure 52: The sphere-of-influence hull (heavy contour) of the points in Fig. 51.

Figure 53: Another set of points.

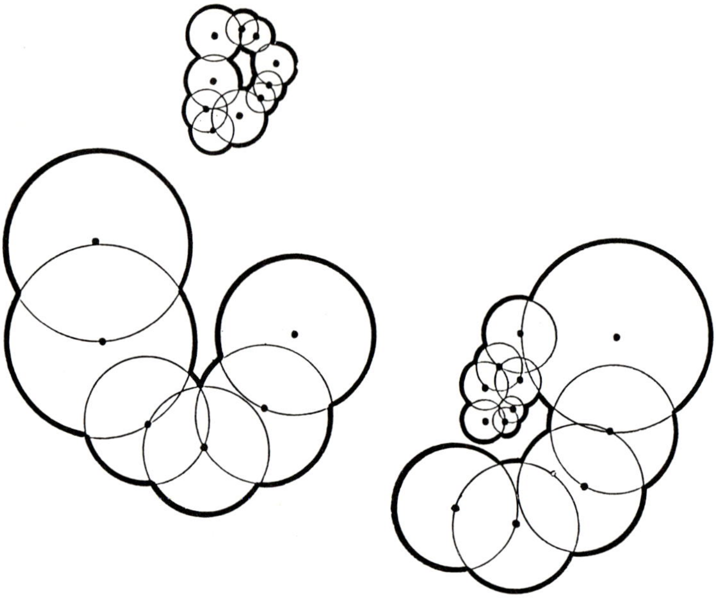

Figure 54: The sphere-of-influence hull (heavy contour) of the points in Fig. 53.

AUTHOR INDEX